CAMBRIDGE LIBRARY COLLECTION

Books of enduring scholarly value

Physical Sciences

From ancient times, humans have tried to understand the workings of the world around them. The roots of modern physical science go back to the very earliest mechanical devices such as levers and rollers, the mixing of paints and dyes, and the importance of the heavenly bodies in early religious observance and navigation. The physical sciences as we know them today began to emerge as independent academic subjects during the early modern period, in the work of Newton and other 'natural philosophers', and numerous sub-disciplines developed during the centuries that followed. This part of the Cambridge Library Collection is devoted to landmark publications in this area which will be of interest to historians of science concerned with individual scientists, particular discoveries, and advances in scientific method, or with the establishment and development of scientific institutions around the world.

The Herschels and Modern Astronomy

The Herschels in this biography are Sir William Herschel (1738–1822), his sister Caroline (1750–1848) and Sir John Herschel (1792–1871), William's son. Sir William was an astronomer and telescope-maker who discovered the planet Uranus in 1781. He was appointed 'the King's astronomer' to George III in 1782, and under his patronage built the then largest telescope in the world. Caroline Herschel worked as her brother's assistant for much of his career but was also an accomplished astronomer in her own right, discovering eight comets and producing a catalogue of nebulae. Her nephew Sir John Herschel was also a distinguished astronomer who made many observations of stars in the southern hemisphere. This book by the astronomer and writer Agnes Clerke (1842–1907), published in 1895, provides both an analysis of their work and an assessment of its contribution to later astronomical research.

Cambridge University Press has long been a pioneer in the reissuing of out-of-print titles from its own backlist, producing digital reprints of books that are still sought after by scholars and students but could not be reprinted economically using traditional technology. The Cambridge Library Collection extends this activity to a wider range of books which are still of importance to researchers and professionals, either for the source material they contain, or as landmarks in the history of their academic discipline.

Drawing from the world-renowned collections in the Cambridge University Library, and guided by the advice of experts in each subject area, Cambridge University Press is using state-of-the-art scanning machines in its own Printing House to capture the content of each book selected for inclusion. The files are processed to give a consistently clear, crisp image, and the books finished to the high quality standard for which the Press is recognised around the world. The latest print-on-demand technology ensures that the books will remain available indefinitely, and that orders for single or multiple copies can quickly be supplied.

The Cambridge Library Collection will bring back to life books of enduring scholarly value (including out-of-copyright works originally issued by other publishers) across a wide range of disciplines in the humanities and social sciences and in science and technology.

The Herschels and
Modern Astronomy

AGNES MARY CLERKE

CAMBRIDGE
UNIVERSITY PRESS

CAMBRIDGE UNIVERSITY PRESS

Cambridge, New York, Melbourne, Madrid, Cape Town, Singapore,
São Paolo, Delhi, Dubai, Tokyo

Published in the United States of America by Cambridge University Press, New York

www.cambridge.org
Information on this title: www.cambridge.org/9781108013925

© in this compilation Cambridge University Press 2010

This edition first published 1895
This digitally printed version 2010

ISBN 978-1-108-01392-5 Paperback

THE CENTURY SCIENCE SERIES.

EDITED BY SIR HENRY E. ROSCOE, D.C.L., LL.D., F.R.S.

THE HERSCHELS

AND

MODERN ASTRONOMY

SIR WILLIAM HERSCHEL.
Ætat. 50.

(From Abbott's painting in the National Portrait Gallery.)

THE CENTURY SCIENCE SERIES.

THE HERSCHELS

AND

MODERN ASTRONOMY

BY

AGNES M. CLERKE

AUTHOR OF
"A POPULAR HISTORY OF ASTRONOMY DURING THE 19TH CENTURY,"
"THE SYSTEM OF THE STARS," ETC.

CASSELL AND COMPANY, LIMITED

LONDON, PARIS & MELBOURNE

1895

PREFACE.

THE chief authority for the Life of Sir William
Herschel is Mrs. John Herschel's "Memoir of Caroline
Herschel" (London, 1876). It embodies Caroline's
Journals and Recollections, the accuracy of which
is above suspicion. William himself, indeed, referred
to her for dates connected with his early life. The
collateral sources of information are few and meagre:
they yield mere gleanings, yet gleanings worth collecting. Professor E. S. Holden has had recourse to
many of them for his excellent little monograph
entitled "Herschel, his Life and Works" (London,
1881), which is usefully supplemented by "A Synopsis
of the Scientific Writings of Sir William Herschel,"
prepared by the same author with the aid of Professor Hastings. It made part of the Smithsonian
Report for 1880, and was printed separately at Wash-
ington in 1881. But the wonderful series of papers it
summarises have still to be sought, one by one, by
those desiring to study them effectually, in the various
volumes of the *Philosophical Transactions* in which
they originally appeared. Their collection and re-
publication is, nevertheless, a recognised desideratum,
and would fill a conspicuous gap in scientific literature.

Sir John Herschel's life has yet to be written. The published materials for it are scanty, although they have been reinforced by the inclusion in the late Mr. Graves's " Life of Sir William Rowan Hamilton " (Dublin, 1882-9) of his correspondence with that remarkable man. The present writer has, however, been favoured by the late Miss Herschel, and by Sir William J. Herschel, with the perusal of a considerable number of Sir John Herschel's, as well as of Sir William's, manuscript letters. She also gratefully acknowledges the kind help afforded to her by Lady Gordon and Miss Herschel in connection with the portraits reproduced in this volume. For detailed bibliographical references, the articles on Sir John, Sir William, and Caroline Herschel, in the " Dictionary of National Biography," may be consulted.

CONTENTS.

———●◇●———

ILLUSTRATIONS.

THE HERSCHELS

AND

MODERN ASTRONOMY.

———•◇•———

CHAPTER I.

EARLY LIFE OF WILLIAM HERSCHEL.

WILLIAM HERSCHEL was descended from one of three brothers, whose Lutheran opinions made it expedient for them to quit Moravia early in the seventeenth century. Hans Herschel thereupon settled as a brewer at Pirna, in Saxony; his son Abraham rose to some repute as a landscape-gardener in the royal service at Dresden; and Abraham's youngest son, Isaac, brought into the world with him, in 1707, an irresistible instinct and aptitude for music. Having studied at Berlin, he made his way in 1731 to Hanover, where he was immediately appointed oboist in the band of the Hanoverian Guard. A year later he married Anna Ilse Moritzen, by whom he had ten children. The fourth of these, Frederick William, known to fame as *William* Herschel, was born November 15th, 1738.

His brilliant faculties quickly displayed themselves. At the garrison-school he easily distanced his brother Jacob, his senior by four years, and learned besides, privately, whatever French and mathematics

B

the master could teach him.   He showed also a pronounced talent for music, and was already, at fourteen,
a proficient on the hautboy and violin.   In this direction lay his manifest destiny.   His father was now band-
master of the Guard; he was poor, and had no other
provision to give his sons than to train them in his
own art; and thus William, driven by necessity to
become self-supporting while still a boy, entered the
band as oboist in 1753.   They were a family of
musicians.   Of the six who reached maturity, only
Mrs. Griesbach, the elder daughter, gave no sign of
personally owning a share in the common gift, which
descended, nevertheless, to her five sons, all noted
performers on sundry instruments.

William Herschel accompanied his regiment to
England in 1755, with his father and elder brother.
He returned a year later, bringing with him a copy of
Locke " On the Human Understanding," upon which
he had spent the whole of his small savings.   Two
of the three volumes thus acquired were recovered
by his sister after seventy years, and transmitted to
his son   The breaking-out of the Seven Years' War
proved decisive as to his future life.   Campaigning
hardships visibly told upon his health; his parents
resolved, at all hazards, to rescue him from them;
and accordingly, after the disaster at Hastenbeck,
July 26th, 1757, they surreptitiously shipped him off
to England.   By this adventure, since he was in the
military service of the Elector of Hanover, George III.
of England, he incurred the penalties of desertion;
but they were never exacted, and were remitted by
the King himself in 1782.

William Herschel was in his nineteenth year
when he landed at Dover with a French crown-piece

in his pocket. Necessity or prudence kept him for some time obscure; and we next hear of him as having played a solo on the violin at one of Barbandt's concerts in London, February 15th, 1760. In the same year he was engaged by the Earl of Darlington to train the band of the Durham Militia, when his shining qualities brought him to the front. The officers of the regiment looked with astonishment on the phenomenal young German who had dropped among them from some cloudy region; who spoke English perfectly, played like a virtuoso, and possessed a curious stock of varied knowledge. Their account of him at a mess-dinner excited the curiosity of Dr. Miller, organist and historian of Doncaster, who, having heard him perform a violin solo by Giardini, fell into a rapture, and invited him on the spot to live with him.

He left nothing undone for the advancement of his *protégé;* procured for him tuitions and leading concert engagements; and encouraged him, in 1765, to compete for the post of organist at Halifax. Herschel's special qualifications were small; his chief rival, Dr. Wainwright, was a skilled player, and at the trial performance evoked much applause by his brilliant execution. Only the builder of the organ, an odd old German named Schnetzler, showed dissatisfaction, exclaiming: "He run about the keys like one cat; he gif my pipes no time for to shpeak." Then Herschel mounted the loft, and the church was filled with a majestic volume of sound, under cover of which a stately melody made itself heard. The "Old Hundredth" followed, with equal effect. Schnetzler was beside himself with delight. "I vil luf dis man," he cried, " because he gif my pipes time for to

B 2

shpeak." Herschel had virtually provided himself
with four hands. A pair of leaden weights brought
in his pocket served to keep down two keys an octave
apart, while he improvised a slow air to suit the con-
tinuous bass thus mechanically supplied. The artifice
secured him the victory.

This anecdote is certainly authentic. It is related
by Dr. Miller from personal knowledge. Nor is it
inconsistent with a story told by Joah Bates, of
King's College, Cambridge, a passionate lover of
music. Repairing to Halifax, his native place, to
conduct the "Messiah" at the opening of a new
organ, he was accosted in the church by a young
man, who asked for an opportunity of practising
on it. Although as yet, he said, unacquainted
with the instrument, he aspired to the place of
organist; and the absolute certitude of his manner so
impressed Bates that he not only granted his request,
but became his warm patron. The young man's
name was William Herschel. We hear, further, on
Dr. Burney's authority, that he played first violin in
Bates's orchestra.

But the tide of his fortunes was flowing, and he
knew how to "take it at the flood." Early in 1766
he removed to Bath as oboist in Linley's celebrated
orchestra, which played daily in the Pump Room to
enliven the parade of blushing damsels and ruffling
gallants pictured to our fancy in Miss Austen's novels.
Bath was then what Beau Nash had made it—the
very focus of polite society. Turbans nodded over
cards; gigs threaded their way along Union Passage;
Cheap Street was blocked with vehicles; the Lower
Rooms witnessed the nightly evolutions of the
country-dance; the Grove, as Doran reminds us, was

brilliant with beauty, coquelicot ribbons, smart pel-
isses, laced coats, and ninepins. The feat of "tipping
all nine for a guinea" was frequently performed;
and further excitement might be had by merely
plucking some lampoons from the trees, which seemed
to bear them as their natural fruit. Music, too, was
in high vogue. The theatres were thronged; and
Miss Linley's exquisite voice was still heard in the
concert-halls.

On the 4th of October, 1767, the new Octagon
Chapel was opened for service, with Herschel as
organist. How it was that he obtained this "agree-
able and lucrative situation" we are ignorant; but
he had that singular capacity for distinction which
explains everything. The Octagon Chapel became a
centre of fashionable attraction, and he soon found
himself lifted on the wave of public favour. Pupils
of high rank thronged to him, and his lessons
often mounted to thirty-five a week. He composed anthems, psalm-tunes, even full services for
his assiduously-trained choir. His family were made
sharers in his success. He secured a post in Linley's
orchestra for his younger brother Alexander, in 1771;
and he himself fetched his sister Caroline to Bath in
1772. Both were of very considerable help to him in
his musical and other enterprises, the latter of which
gradually gained ground over the former.

Music was never everything to William Herschel.
He cultivated it with ardour; composed with facility
in the prevalent graceful Italian style; possessed a
keen appreciation and perfect taste. But a musical
career, however brilliant, did not satisfy him. The
inner promptings of genius told him to look beyond.
The first thirty-five years of his life were thus spent

in diligently preparing to respond to an undeclared vocation. Nothing diverted him from his purpose of self-improvement. At first, he aimed chiefly at mastering the knowledge connected with his profession. With a view to the theory of music, " I applied myself early," he said, in a slight autobiographical sketch sent to Lichtenberg at Göttingen, " to all the branches of the mathematics, algebra, conic sections, fluxions, etc. Contracting thereby an insatiable desire for knowledge in general, I extended my application to languages—French, Italian, Latin, English—and determined to devote myself entirely to the pursuit of knowledge, in which I resolved to place all my future enjoyment and felicity. This resolution I have never had occasion to change." At Bath, in the midst of engrossing musical occupations, his zeal for study grew only the more intense. After fourteen or sixteen hours of teaching, he would "unbend his mind" by plunging into Maclaurin's " Fluxions," or retire to rest with a basin of milk, Smith's " Opticks," and Ferguson's "Astronomy." He had no sooner fallen under the spell of this last science than he " resolved to take nothing upon trust, but to see with my own eyes all that other men had seen before."

He hired, to begin with, a small reflector; but what it showed him merely whetted his curiosity. And the price of a considerably larger instrument proved to be more than he could afford to pay. Whereupon he took the momentous resolution of being, for the future, his own optician. This was in 1772. He at first tried fitting lenses into pasteboard tubes, with the poor results that can be imagined. Then he bought from a Quaker, who had dabbled in that line, the discarded rubbish of his tools, patterns, polishers, and abortive mirrors; and in June, 1773, when fine folk had mostly deserted Bath for summer resorts, work was begun in earnest. The house was turned topsy-turvy; the two brothers attacked the novel enterprise with boyish glee. Alexander, a born mechanician, set up a huge lathe in one of the bed-rooms; a cabinet-maker was installed in the drawing-room; Caroline, in spite of secret dismay at such unruly proceedings, lent a hand, and kept meals going; William directed, inspired, toiled, with the ardour of a man who had staked his life on the issue. Meanwhile, music could not be neglected. Practising and choir-training went on; novelties for the ensuing season were prepared; compositions written, and parts copied. Then the winter brought the usual round of tuitions and performances, while all the time mirrors were being ground and polished, tried and rejected, without intermission. At last, after *two hundred* failures, a tolerable reflecting telescope was produced, about five inches in aperture, and of five and a half feet focal length. The outcome may seem small for so great an expenditure of pains; but those two hundred failures made the Octagon Chapel organist an expert, unapproached and unapproach-able, in the construction of specula. With his new instrument, on March 4th, 1774, he observed the Nebula in Orion; and a record of this beginning of his astronomical work is still preserved by the Royal Society.

William Herschel was now, as to age, *in mezzo cammin.* He had numbered just so many years as had Dante when he began the "Divina Commedia." But he had not, like Dante, been thrown off the rails of life. The rush of a successful professional career was irresistibly carrying him along. Almost any other man would have had all his faculties absorbed in it. Herschel's were only stimulated by the occupations which it brought. Yet they were of a peculiarly absorbing nature. Music is the most exclusive of arts. In turning aside, after half a lifetime spent in its cultivation, to seek his ideal elsewhere, Herschel took an unparalleled course. And his choice was final. Music was long his pursuit, astronomy his pastime; a fortunate event enabled him to make astronomy his pursuit, while keeping music for a pastime.

Yet each demands a totally different kind of training, not only of the intellect, but of the senses. From his earliest childhood William Herschel's nerves and brain had been specially educated to discriminate impressions of sound, and his muscles to the peculiar agility needed for their regulated and delicate production; while, up to the age of thirty-five, he had used his eyes no more purposefully than other people. The eye, nevertheless, requires cultivation as much as the ear. "You must not expect to *see at sight*," he told Alexander Aubert, of Loam Pit Hill, in 1782. And he wrote to Sir William Watson: "Seeing is in some respects an art which must be learnt. Many a night have I been practising to see, and it would be strange if one did not acquire a certain dexterity by such constant practice." A critical observation, he added, could no more be expected from a novice at the telescope than a performance of one of Handel's organ-fugues from a beginner in music. In this difficult art of vision he rapidly became an adept. Taking into account the full extent of his powers, the opinion has been expressed, and can scarcely be contradicted, that he never had an equal.

At midsummer, 1774, Herschel removed from No. 7, New King Street, to a house situated near Walcot Turnpike, Bath. A grass-plot was attached to the new residence, and it afforded convenient space for workshops. For already he designed to " carry improvements in telescopes to their utmost extent," and " to leave no spot of the heavens unvisited." An unprecedented ambition ! No son of Adam had ever before entertained the .like. To search into the recesses of space, to sound its depths, to dredge up from them their shining contents, to classify these, to investigate their nature, and trace their mutual relations, was what he proposed to do, having first provided the requisite optical means. All this in the intervals of professional toils, with no resources except those supplied by his genius and ardour, with no experience beyond that painfully gained during the progress of his gigantic task.

Since the time of Huygens, no systematic attempt had been made to add to the power of the telescope. For the study of the planetary surfaces, upon which he and his contemporaries were mainly intent, such addition was highly desirable. But Newton's discovery profoundly modified the aims of astronomers. Their essential business then became that of perfecting the theories of the heavenly bodies. Whether or not they moved in perfect accordance with the law of gravitation was the crucial question of the time. Newton's generalisation was on its trial. Now and again it almost seemed as if about to fail. But difficulties arose only to be overcome, and before the eighteenth century closed the superb mechanism of the planetary system was elucidated. Working flexibly under the control of a single dominant force, it was shown to possess a self-righting power which secured its indefinite duration. Imperishable as the temple of Poseidon, it might be swayed by disturbances, but could not be overthrown.

The two fundamental conclusions—that the Newtonian law is universally valid, and that the solar system is a stable structure—were reached by immense and sustained labours. Their establishment was due, in the main, to the mathematical genius of Clairaut, D'Alembert, Lagrange, and Laplace. But refined analysis demands refined data ; hence the need for increased accuracy of observation grew continually more urgent. Attention was accordingly concentrated upon measuring, with the utmost exactitude, the places at determinate epochs of the heavenly bodies. The one thing needful was to learn the " when " and " where " of each of them—that is, to obtain such information as the transit-instrument is adapted to give. In this way the deviations of the moon and planets from their calculated courses became known; and upon the basis of these " errors " improved theories were built, then again compared with corrected observations.

For these ends, large telescopes would have been useless. They were not, however, those that Herschel had in view. The *nature* of the orbs around us, not their motions, formed the subject of his inquiries, with which modern descriptive astronomy virtually originated. He was, moreover, the founder of sidereal astronomy. The stars had, until his career began, received little *primary* attention. They were regarded and observed simply as reference-points by which to track the movements of planets, comets, and the moon. Indispensable for fiducial purposes, they almost escaped consideration for themselves. They were, indeed, thought to lie beyond the reach of effective investigation. Only the outbursts of temporary stars, and the fluctuations of two or three periodical ones, had roused special interest, and seemed deserving of particular inquiry.

Of the dim objects called "nebulæ," Halley had counted up half a dozen in 1714; Lacaille compiled a list of forty-two at the Cape, in 1752–55; and Messier published at Paris, in 1771, a catalogue of forty-five, enlarged to one hundred and three in 1781. He tabulated, only to rid himself of embarrassments from them. For he was *by trade* a comet-hunter, and, until he hit upon this expedient, had been much harassed in its exercise by mistakes of identity.

But Herschel did not merely "pick up;" he explored. This was what no one before him had thought of doing. A "review of the heavens" was a complete novelty. The magnificence of the idea, which was rooted in his mind from the start, places him apart from, and above, all preceding observers.

To its effective execution telescopic development was essential. The two projects of optical improvement and of sidereal scrutiny went together. The skies could be fathomed, if at all, only by means of light-collecting engines of unexampled power. Rays enfeebled by distance should be rendered effective by concentration. Stratum after stratum of bodies —

"Clusters and beds of worlds, and bee-like swarms
Of suns and starry streams,"

previously unseen, and even unsuspected, might, by the strong focussing of their feebly-surviving rays, be

brought to human cognisance. The contemplated
" reviews" would then be complete just in proportion
to the grasp of the instrument used in making them.

The first was scarcely more than a reconnaissance.
It was made in 1775, with a small reflector of the
Newtonian make.* Its upshot was to impress him
with the utter disproportion between his daring plans
and the means as yet at his disposal. Speculum-
casting accordingly recommenced with fresh vigour.
Seven- and ten-foot mirrors were succeeded by others
of twelve, and even of twenty feet focal length. The
finishing of them was very laborious. It was at that
time a manual process, during the course of which the
hands could not be removed from the metal without
injury to its figure. One stretch of such work lasted
sixteen hours, Miss Herschel meantime, " by way of
keeping him alive," putting occasional morsels of food
into the diligent polisher's mouth. His mode of pro-
cedure was to cast and finish many mirrors of each
sort; then to select the best by trial, and repolish the
remainder. In this manner he made, before 1781,
" not less than 200 seven-foot, 150 ten-foot, and about
80 twenty-foot mirrors, not to mention those of the
Gregorian form." Repolishing operations were, moreover, accompanied by constant improvements, so that
each successive speculum tended to surpass its predecessors.

These absorbing occupations were interrupted by
the unwelcome news that Dietrich, the youngest of

* In "Newtonian" telescopes the image formed by the large
speculum is obliquely reflected from a small plane mirror to the side
of the tube, where it is viewed with an ordinary eye-piece. With a
"Gregorian," the observer looks straight forward, the image being
thrown back by a little *concave* mirror through a central perforation
in the speculum where the eye-piece is fitted.

the Herschel family, had decamped from Hanover
" with a young idler" like himself. William instantly
started for Holland, where the fugitive was supposed
to be about to take ship for India, but missed his
track; and, after having extended his journey to
Hanover to comfort his anxious mother—his father
had died in 1767—returned sadly to Bath. There, to
his immense surprise, he found the scapegrace in
strict charge of his sister, "who kept him to a diet
of roasted apples and barley-water." His ineffectual
escapade had terminated with an attack of illness at
Wapping, whither Alexander Herschel, on learning
how matters stood, had posted off to take him in
charge and watch his recovery. Musical occupation
was easily procured for him at Bath, since he was an
accomplished violinist—had, indeed, started on his
unprosperous career in the guise of an infant prodigy;
but he threw it up in 1779 and drifted back to
Hanover, married a Miss Reif, and settled down to
live out a fairly long term of shiftless, albeit harmless,
existence.

In 1776 William Herschel succeeded Thomas
Linley, Sheridan's father-in-law, as Director of the
Public Concerts at Bath. His duties in this capacity,
while the season lasted, were most onerous. He had
to engage performers, to appease discontents, to
supply casual failures, to write glees and catches
expressly adapted to the voices of his executants,
frequently to come forward himself as a soloist on
the hautboy or the harpsichord. The services of his
brother Alexander, a renowned violoncellist, and of
his sister, by this time an excellent singer, were now
invaluable to him. Nor for musical purposes solely.
The vision of the skies was never lost sight of, and

the struggle to realise it in conjunction with his sympathetic helpers absorbed every remnant of time.   At
meals the only topics of conversation were mechanical
devices for improving success and averting failure.
William ate with a pencil in his hand, and a project
in his head.   Between the acts at the theatre, he
might be seen running from the harpsichord to his
telescope.   After a rehearsal or a morning performance, he would dash off to the workshop in periwig
and lace ruffles, and leave it but too often with
those delicate adjuncts to his attire torn and pitch-
bespattered.   Accidents, too, menacing life and limb,
were a consequence of that "uncommon precipitancy
which accompanied all his actions;" but he escaped
intact, save for the loss of a finger-nail.

His introduction to the learned world of Bath was
thus described by himself:—

"About the latter end of December, 1779, I happened to be
engaged in a series of observations on the lunar mountains;
and the moon being in front of my house, late in the evening
I brought my seven-feet reflector into the street, and directed
it to the object of my observations.   Whilst I was looking into
the telescope, a gentleman, coming by the place where I was
stationed, stopped to look at the instrument.   When I took
my eye off the telescope, he very politely asked if he might be
permitted to look in, and this being immediately conceded, he
expressed great satisfaction at the view."

The inquisitive stranger called next morning, and
proved to be Dr. (later Sir William) Watson.   He
formed on the spot an unalterable friendship for the
moon-struck musician, and introduced him to a
Philosophical Society which held its meetings at his
father's house.   Herschel's earliest essays were read
before it, but they remained unpublished.   His first
printed composition appeared in the " Ladies' Diary "

for 1780. It was an answer to a prize question on the vibration of strings.

The long series of his communications to the Royal Society of London opened May 11th, 1780, with a discussion of his observations, begun in October, 1777, of Mira, the variable star in the neck of the Whale. As to the theory of its changes, he agreed with Keill that they could best be explained by supposing rotation on an axis to bring a lucid side and a side obscured by spots alternately into view. A second paper by him on the Mountains of the Moon was read on the same day. He measured, in all, about one hundred of these peaks and craters.

In January, 1781, there came an essay stamped with the peculiar impress of his genius, entitled " Astronomical Observations on the Rotation of the Planets round their Axes, made with a view to determine whether the earth's diurnal motion is perfectly equable." It embodied an attempt to apply a definite criterion to the time-keeping of our planet. But the prospect is exceedingly remote of rating one planet-clock by the other. Herschel's methods of inquiry are, however, aptly illustrated in this curiously original paper. His speculations always invited the control of facts. If facts were not at hand, he tried somehow to collect them. The untrammelled play of fancy was no more to his mind than it was to Newton's. His ardent scientific imagination was thus, by the sobriety of his reason, effectively enlisted in the cause of progress.

Herschel began in 1780 his second review of the heavens, using a seven-foot Newtonian, of $6\frac{1}{4}$ inches aperture, with a magnifying power of 227. " For distinctness of vision," he said, " this instrument is, perhaps, equal to any that was ever made." His praise was amply justified. As he worked his way with it through the constellation Gemini, on the night of March 13th, 1781, an unprecedented event occurred. "A new planet swam into his ken." He did not recognise it as such. He could only be certain that it was not a fixed star. His keen eye, armed with a perfect telescope, discerned at once that the object had a disc; and the application of higher powers showed the disc to be a substantial reality. The stellar "patines of bright gold" will not stand this test. Being of purely optical production, they gain nothing by magnification.

At that epoch new planets had not yet begun to be found by the dozen. Five, besides the earth, had been known from the remotest antiquity. Five, and no more, seemed to have a prescriptive right to exist. The boundaries of the solar system were of imme-morial establishment. It was scarcely conceivable that they should need to be enlarged. The notion did not occur to Herschel. His discovery was modestly imparted to the Royal Society as "An Account of a Comet." He had, indeed, noticed that the supposed comet moved in planetary fashion from west to east, and very near the ecliptic; and, after a few months, its true nature was virtually proved by Lexell of St. Petersburg. On November 28th, Herschel measured, with his freshly-invented "lamp-micrometer," the diameter of this "singular star;" and it was not until a year later, November 7th, 1782, that he felt suffi-ciently sure of its planetary status to exercise his right of giving it a name. Yet this, in the long run, he failed to accomplish. The appellation " Georgium Sidus," bestowed in honour of his patron, George III.,

never crossed the Channel, and has long since gone
out of fashion amongst ourselves.   Lalande tried
to get the new planet called "Herschel;" but the
title "Uranus," proposed by Bode, of Berlin, was the
"fittest," and survived.

This discovery made the turning-point of Herschel's
career.   It transformed him from a music-master into
an astronomer.   Without it his vast abilities would
probably have been in great measure wasted.   No
man could long have borne the strain of so arduous a
double life as he was then leading.   Relief from it
came just in time.   It is true that fame, being often
more of a hindrance than a help, brought embarrassments in its train.   In November, 1781, Herschel was
compelled to break the complex web of his engagements at Bath by a journey to London for the purpose of receiving in person the Copley Medal awarded
to him by the Royal Society, of which body he was,
some days later, elected a Fellow.   At home, he was
persecuted by admirers; and they were invariably
received with an easy suavity of manner that gave no
hint of preoccupation.   Everyone of scientific pretension who visited Bath sought an interview with the
extraordinary man who, by way of interlude to pressing duties, had built telescopes of unheard-of power,
and performed the startling feat of adding a primary
member to the solar system.   Among the few of
these callers whose names have been preserved were
Sir Harry Englefield, Sir Charles Blagden, and
Dr. Maskelyne, then, and for thirty years afterwards,
Astronomer-Royal.   "With the latter," Miss Herschel
relates, "he (William) was engaged in a long conversation which to me sounded like quarrelling, and the first
words my brother said after he was gone were, 'That

c

is a devil of a fellow!'" The phrase was doubtless meant as a sign of regard, for the acquaintance thus begun ripened into cordial intimacy. And William Herschel never lost or forgot a friend.

As regards music alone, the winter of 1781-82 was an exceptionally busy one. He had arranged to conduct, jointly with Rauzzini, a Roman singer and composer, a series of oratorios; undertaking, besides, pecuniary responsibilities which turned out little to his advantage. The labour, vexation, and disappointment involved in carrying out this unlucky plan can readily be imagined. But neither the pressure of business, nor the distractions of celebrity, checked the ardour of his scientific advance. The review which afforded him the discovery of Uranus, and the materials for his first catalogue of 269 double stars, was completed in 1781; and a third, made with the same beautiful instrument, bearing the high magnifying power of 460, was promptly begun. This had for one of its special objects the ascertainment of possible changes in the heavens since Flamsteed's time; and in the course of it many thousands of stars came under scrutiny, directed to ascertain their magnitude and colour, singleness or duplicity, hazy or defined aspect.

The first of Herschel's *effective* twenty-foot telescopes was erected at 19, New King Street, in the summer of 1781. Enclosing a mirror twelve inches in diameter, it far surpassed any seeing-machine that had ever existed in the world. Yet its maker regarded it as only marking a step in his upward progress. A speculum of thirty-feet focus was the next object of his ambition. For its achievement no amount of exertion was counted too great. Its composition was

regulated by fresh experiments on various alloys of
copper and tin. Its weight and shape were again and
again calculated, and the methods appropriate to its
production earnestly discussed.   " I saw nothing else,"
Caroline Herschel tells us, " and heard nothing else
talked of but these things when my brothers were
together."*

" The mirror," she continues, " was to be cast in a
mould of loam prepared from horse-dung, of which an
immense quantity was to be pounded in a mortar and
sifted through a fine sieve. It was an endless piece
of work, and served me for many an hour's exercise ;
and Alex frequently took his turn at it, for we were
all eager to do something towards the great undertaking. Even Sir William Watson would sometimes
take the pestle from me when he found me in the
work-room."

The matter was never out of the master's thoughts.
" If a minute could but be spared in going from one
scholar to another, or giving one the slip, he called at
home to see how the men went on with the furnace,
which was built in a room below, even with the
garden."

At last, the concert season being over, and everything in readiness for the operation of casting, " the
metal," we hear from the same deeply-interested eye-
witness, " was in the furnace ; but, unfortunately, it
began to leak at the moment when ready for pouring,
and both my brothers, and the caster with his men,
were obliged to run out at opposite doors, for the

* In borrowing Miss Herschel's lively narratives and comments,
some obvious slips in grammar and construction have been corrected.
Quotations, too, from the writings of Sir William and Sir John
Herschel are often slightly abridged.

stone flooring, which ought to have been taken up, flew about in all directions, as high as the ceiling. My poor brother William fell, exhausted with heat and exertion, on a heap of brickbats. Before the second casting was attempted, everything which could ensure success had been attended to, and a very perfect metal was found in the mould, which had cracked in the cooling."

This second failure terminated the enterprise. Not that it was abandoned as hopeless, but because of a total change in the current of affairs. Herschel's fame had stirred the royal curiosity, and rumours had now and again reached Bath that he was to be sent for to court. In the spring of 1782 the actual mandate arrived; and on May 8th, leaving his pupils and his projects to shift for themselves, he set out for London. He carried with him his favourite seven-foot reflector, and all the apparatus necessary for viewing double stars and other objects of interest. On May 25th he wrote to his sister :—

"I have had an audience of His Majesty this morning, and met with a very gracious reception. I presented him with the drawing of the solar system, and had the honour of explaining it to him and the Queen. My telescope is in three weeks' time to go to Richmond, and meanwhile to be put up at Greenwich. . . . Tell Alexander that everything looks very like as if I were to stay here. The King enquired after him, and after my great speculum. He also gave me leave to come and hear the Griesbachs (Herschel's nephews) play at the private concert which he has every evening. . . . All my papers are printing, and are allowed to be very valuable. You see, Lina, I tell you all these things.

You know vanity is not my foible, therefore I need not fear your censure. Farewell."

His next letter is dated June 3rd, 1782. " I pass my time," he informed " Lina," " between Greenwich and London agreeably enough, but am rather at a loss for work that I like. Company is not always pleasing, and I would much rather be polishing a speculum. Last Friday I was at the King's concert to hear George play. The King spoke to me as soon as he saw me, and kept me in conversation for half an hour. He asked George to play a solo-concerto on purpose that I might hear him. . . . I am introduced to the best company. To-morrow I dine at Lord Palmerston's, next day with Sir Joseph Banks, etc. Among opticians and astronomers nothing now is talked of but *what they call* my great discoveries. Alas ! this shows how far they are behind, when such trifles as I have seen and done are called *great*. Let me but get at it again ! I will make such telescopes and see such things—that is, I will endeavour to do so."

A comparison of his telescope with those at the Royal Observatory showed its striking superiority, although among them was one of Short's famous Gregorians, of 9½ inches aperture. It had thus a reflecting surface above twice that of Herschel's seven-foot, the competition with which was nevertheless so disastrous to its reputation that Dr. Maskelyne fell quite out of conceit with it, and doubted whether it *deserved* the new stand constructed for it on the model of Herschel's.

In the midst of these scientific particulars, we hear incidentally that influenza was then so rife in London that " hardly one single person " escaped an attack.

On July 2nd he made his first appearance as showman of the heavens to royalty. The scene of the display was Buckingham House (now Buckingham Palace). "It was a very fine evening," he wrote to his sister. "My instrument gave general satisfaction. The King has very good eyes, and enjoys observations with telescopes exceedingly."

Next night, the King and Queen being absent at Kew, the Princesses desired an exhibition. But, since they objected to damp grass, the telescope, Herschel says, "was moved into the Queen's apartments, and we waited some time in hopes of seeing Jupiter or Saturn. Meanwhile I showed the Princesses and several other ladies the speculum, the micrometers, the movements of the telescope, and other things that seemed to excite their curiosity. When the evening appeared to be totally unpromising, I proposed an artificial Saturn as an object, since we could not have the real one. I had beforehand prepared this little piece, as I guessed by the appearance of the weather in the afternoon we should have no stars to look at. This being accepted with great pleasure, I had the lamps lighted up, which illuminated the picture of a Saturn (cut out in pasteboard) at the bottom of the garden wall. The effect was fine, and so natural that the best astronomer might have been deceived. Their royal highnesses seemed to be much pleased with the artifice." From a somewhat prolonged conversation, he judged them to be "extremely well instructed," and "most amiable characters."

Shortly afterwards Herschel received the appointment of royal astronomer, with the modest salary of £200 a year. "Never," exclaimed Sir William Watson on being made acquainted with its amount, " bought

monarch honour so cheap!" The provision was
assuredly not munificent; yet it sufficed to rescue a
great man from submergence under the hard necessities of existence. The offer was critically timed. It
was made precisely when teaching and concert-giving
had come to appear an "intolerable waste of time" to
one fired with a visionary passion. "Stout Cortes"
staring at the Pacific, Ulysses starting from Ithaca to
"sail beyond the sunset," were not more eager for
experience of the Unknown.

CHAPTER II.

THE KING'S ASTRONOMER.

WILLIAM HERSCHEL was now an appendage to the
court of George III. He had to live near Windsor, and
a large dilapidated house on Datchet Common was
secured as likely to meet his unusual requirements.
The "flitting" took place August 1, 1782. William was
in the highest spirits. There were stables available
for workrooms and furnaces ; a spacious laundry that
could be turned into a library; a fine lawn for the
accommodation of the great reflector. Crumbling
walls and holes in the roof gave him little or no
concern ; and if butcher's meat was appallingly dear
(as his sister lamented) the family could live on bacon
and eggs ! In this sunny spirit he entered upon the
career of untold possibilities that lay before him.

Nevertheless the King's astronomer did not find it
all plain sailing. His primary duty was to gratify the
royal taste for astronomy, and this involved no trifling
expenditure of time and toil. The transport of the
seven-foot to the Queen's lodge could be managed in
the daylight, but its return-journey in the dark, after
the conclusion of the celestial raree-show, was an
expensive and a risky business ; yet fetched back it
should be unless a clear night were to be wasted—a
thing not possible to contemplate. This kind of
attendance was, however, considerately dispensed with
when its troublesome nature came to be fully understood. Herschel's treatment by George III. has often

been condemned as selfish and niggardly; but with scant justice. In some respects, no doubt, it might advantageously have been modified. Still, the fact remains that the astronomer of Slough was the gift to science of the poor mad King. From no other crowned head has it ever received so incomparable an endowment.

Herschel's salary was undeniably small. It gave him the means of living, but not of observing, as he proposed to observe. If the improvement of telescopes were to be "carried to its utmost limit," additional funds must be raised. Without an ample supply of the "sinews of war," fresh campaigns of exploration were out of the question. There was one obvious way in which they could be provided. Herschel's fame as an optician was spread throughout Europe. His telescopes were wanted everywhere, but could be had from himself alone; for the methods by which he wrought specula to a perfect figure are even now undivulged. They constituted, therefore, a source of profit upon which he could draw to almost any extent. He applied himself, accordingly, to make telescopes for sale. They brought in large sums. Six hundred guineas a-piece were paid to him by the King for four ten-foot reflectors; he received at a later date £3,150 for a twenty-five foot, sent to Spain; and in 1814 £2,310 from Lucien Bonaparte for two smaller instruments. The regular scale of prices (later considerably reduced) began with 200 guineas for a seven-foot, and mounted to 2,500 for a twenty-foot; and the commissions executed were innumerable.

But Herschel did not come into the world to drive a lucrative trade. It was undertaken, not for itself, but for what was to come of it; yet there was danger lest the end should be indefinitely postponed in the endeavour to secure the means.

"It seemed to be supposed," Miss Herschel remarked, "that enough had been done when my brother was enabled to leave his profession that he might have time to make and sell telescopes. But all this was only retarding the work of a thirty or forty-foot instrument, which it was his chief object to obtain as soon as possible; for he was then on the wrong side of forty-five, and felt how great an injustice he would be doing to himself and the cause of astronomy by giving up his time to making telescopes for other observers."

This he was, fortunately, not long obliged to do. A royal grant of £2,000 for the construction of the designed giant telescope, followed by another of equal amount, together with an annual allowance of £200 for its repairs, removed the last obstacle to his success. The wide distribution of first-class instruments might, indeed, have been thought to promise more for the advancement of astronomy than the labours of a single individual. No mistake could be greater. Not an observation worth mentioning was made with any of the numerous instruments sent out from Datchet or Slough, save only those acquired by Schröter and Pond. The rest either rusted idly, or were employed ineffectually, aptly illustrating the saying that "the man at the eye-end" is the truly essential part of a telescope.

No one knew this better than Herschel. Every serene dark night was to him a precious opportunity availed of to the last minute. The thermometer might descend below zero, ink might freeze, mirrors might crack; but, provided the stars shone, he and his sister worked on from dusk to dawn. In this way, his "third review," begun at Bath, was finished in the spring of 1783. The swiftness with which it was conducted implied no want of thoroughness. "Many a night," he states, "in the course of eleven or twelve hours of observation, I have carefully and singly examined not less than 400 celestial objects, besides taking measures, and sometimes viewing a particular star for half an hour together, with all the various powers."

The assiduity appears well-nigh incredible with which he gathered in an abundant harvest of nebulæ and double stars; his elaborate papers, brimful of invention and experience, being written by day, or during nights unpropitious for star-gazing. On one occasion he is said to have worked without intermission at the telescope and the desk for seventy-two hours, and then slept unbrokenly for twenty-six hours. His instruments were never allowed to remain disabled. They were kept, like himself, on the alert. Relays of specula were provided, and one was in no case removed from the tube for re-polishing, unless another was ready to take its place. Even the meetings of the Royal Society were attended only when moonlight effaced the delicate objects of his particular search.

The summer of 1788 was spent in getting ready the finest telescope Herschel had yet employed. It was called the "large twenty-foot" because of the size of its speculum, which was nearly nineteen inches in diameter; and with its potent help he executed his fourth and last celestial survey. His impatience to begin led him into perilous situations.

"My brother," says Miss Herschel, "began his series of sweeps when the instrument was yet in a

very unfinished state ; and my feelings were not very
comfortable when every moment I was alarmed by a
crack or fall, knowing him to be elevated fifteen feet
or more on a temporary cross-beam instead of a safe
gallery. The ladders had not even their braces at the
bottom ; and one night, in a very high wind, he had
hardly touched the ground before the whole apparatus
came down. Some labouring men were called up to
help in extricating the mirror, which was fortunately
uninjured, but much work was cut out for carpenters
next day."

In the following March, he himself wrote to Patrick
Wilson, of Glasgow, son of Dr. Alexander Wilson, the
well-known professor of astronomy :—" I have finished
a second speculum to my new twenty-foot, very much
superior to the first, and am now reviewing the heavens
with it. This will be a work of some years ; but it is
to me so far from laborious that it is attended with
the utmost delight." He, nevertheless, looked upon
telescopes as " yet in their infant state."

The ruinous mansion at Datchet having become
uninhabitable, even by astronomers, their establishment was shifted, in June, 1785, to Clay Hall, near
Old Windsor. Here the long-thought-of forty-foot was
begun, but was not destined to be finished. A litigious
landlady intervened. The next move, however, proved
to be the last. It was to a commodious residence at
Slough, now called " Observatory House "—" le lieu du
monde," wrote Arago, "où il a été fait le plus de
découvertes." Thither, without the loss of an hour, in
April, 1786, the machinery and apparatus collected at
Clay Hall were transported. Yet, " amidst all this
hurrying business," Caroline remembered " that every
moment after daylight was allotted to observing. The

last night at Clay Hall was spent in sweeping till day-light, and by the next evening the telescope stood ready for observation at Slough."

During the ensuing three months, thirty to forty workmen were constantly employed, "some in felling and rooting out trees, some in digging and preparing the ground for the bricklayers who were laying the foundation for the telescope." "A whole troop of labourers" were, besides, engaged in reducing "the iron tools to a proper shape for the mirror to be ground upon." Thus, each morning, when dawn compelled Herschel to desist from observation, he found a bevy of people awaiting instructions of all sorts from him. "If it had not been," his sister says, "for the intervention of a cloudy or moonlit night, I know not when he, or I either, should have got any sleep." The wash-house was turned into a forge for the manufacture of specially designed tools; heavy articles cast in London were brought by water to Windsor; the library was so encumbered with stores, models, and implements, that "no room for a desk or an atlas remained."

On July 3rd, 1786, Herschel, accompanied by his brother Alexander, started for Göttingen, commissioned by the King to present to the University one of the ten-foot reflectors purchased from him. He was elected a Member of the Royal Society of Göttingen, and spent three weeks at Hanover with his aged mother, whom he never saw again. During his absence, however, the forty-foot progressed in accord-ance with the directions he had taken care to leave behind. He trusted nothing to chance. "There is not one screwbolt," his sister asserted, "about the whole apparatus but what was fixed under the immediate eye of my brother. I have seen him lie stretched many an hour in a burning sun, across the top beam, whilst the iron-work for the various motions was being fixed. At one time no less than twenty-four men (twelve and twelve relieving each other) kept polishing day and night; my brother, of course, never leaving them all the while, taking his food without allowing himself time to sit down to table."

At this stage of the undertaking it became the fashion with visitors to use the empty tube as a promenade. Dr. and Miss Burney called, in July, 1786, " to see, and *take a walk* through the immense new telescope." " It held me quite upright," the authoress of " Evelina " related, " and without the least inconvenience; so would it have done had I been dressed in feathers and a bell-hoop."

George III. and the Archbishop of Canterbury followed the general example; and the prelate being incommoded by the darkness and the uncertain footing, the King, who was in front, turned back to help him, saying: " Come, my lord bishop, I will show you the way to heaven." On another occasion " God save the King " was sung and played within the tube by a large body of musicians; and the rumour went abroad that it had been turned into a ball-room !

The University of Oxford conferred upon Herschel, in 1786, an honorary degree of LL.D.; but he cared little for such distinctions. Miss Burney characterised him as a " man without a wish that has its object in the terrestrial globe; " the King had " not a happier subject." The royal bounty, she went on " enables him to put into execution all his wonderful projects, from which his expectations of future discoveries are so sanguine as to make his present existence a state of almost perfect enjoyment." Nor was it possible to "admire his genius more than his gentleness." Again, after taking tea in his company in the Queen's lodge: "this very extraordinary man has not more fame to awaken curiosity than sense and modesty to gratify it. He is perfectly unassuming, yet openly happy; and happy in the success of those studies which would render a mind less excellently formed presumptuous and arrogant." Mrs. Papendick, another court chronicler, says that "he was fascinating in his manner, and possessed a natural politeness, and the abilities of a superior nature."

His great telescope took rank, before and after its completion, as the chief scientific wonder of the age. Slough was crowded with sightseers. All the ruck of Grand Dukes and Serene Highnesses from abroad, besides royal, noble, and gentle folk at home, flocked to gaze at it and interrogate its maker with ignorant or intelligent wonder. The Prince of Orange was a particularly lively inquirer. On one of his calls at Slough, about ten years after the erection of the forty-foot, finding the house vacant, he left a memorandum asking if it were true, as the newspapers reported, that " Mr. Herschel had discovered a new star whose light was not as that of the common stars, but with swallow-tails, as stars in embroidery ? " !

Pilgrim-astronomers came, too—Cassini, Lalande, Méchain and Legendre from Paris, Oriani from Milan, Piazzi from Palermo. Sniadecki, director of the observatory of Cracow, " took lodgings," Miss Herschel relates, " in Slough, for the purpose of seeing and hearing my brother whenever he could find him at leisure. He was a very silent man." One cannot

help fearing that he was also a very great bore. Von
Magellan, another eminent foreign astronomer, communicated to Bode an interesting account of Herschel's
methods of observation. The multitude of entries in
his books astonished him. In sweeping, he reported,
" he lets each star pass at least three times through
the field of his telescope, so that it is impossible that
anything can escape him." The thermometer in the
garden stood that night, January 6th, 1785, at 13 deg.
Fahrenheit; but the royal astronomer, his visitor remarked, "has an excellent constitution, and thinks
about nothing else in the world but the celestial bodies."

In January, 1787, Herschel made trial with his
twenty-foot reflector of the " front-view " plan of construction, suggested by Lemaire in 1732, but never
before practically tested. All that had to be done was
to remove the small mirror, and slightly *tilt* the large
one. The image was then formed close to the upper
margin of the tube, into which the observer, turning
his back to the heavens, looked down. The purpose
of the arrangement was to save the light lost in the
second reflection; and its advantage was at once illustrated by the discovery of two Uranian moons—one
(Titania) circling round its primary in about $8\frac{3}{4}$ hours,
the other (Oberon) in $13\frac{1}{2}$ hours. In order to assure
these conclusions, he made a sketch beforehand of
what *ought* to be seen on February 10th; and on that
night, to his intense satisfaction, "the heavens," as he
informed the Royal Society, "displayed the original
of my drawing by showing, in the situation I had
delineated them, *the Georgian planet attended by two
satellites.* I confess that this scene appeared to me
with additional beauty, the little secondary planets
seeming to give a dignity to the primary one which

raises it into a more conspicuous situation amongt he great bodies of our solar system."

This brilliant result determined him to make a "front-view" of the forty-foot. Its advance towards completion was not without vicissitudes. The first speculum, when put into the tube, February 19th 1787, was found too thin to maintain its shape. A second, cast early in 1788, cracked in cooling. The same metal having been recast February 16th, the artist tried it upon 'Saturn in October; but the effect disappointing his expectation, he wrought at it for ten months longer. At last, after a few days' polishing with his new machine, he turned the great speculum towards Windsor Castle; when its high quality became at once manifest. And such was his impatience to make with it a crucial experiment, that—as he told Sir Joseph Banks—he directed it to the heavens (August 28th, 1789) before it had half come to its proper lustre. The stars came out well, and no sooner had he got hold of Saturn than a sixth satellite stood revealed to view! Its "younger brother" was detected September 17th; and the two could be seen, on favourable opportunities, threading their way, like beads of light, along the lucid line of the almost vanished ring. Herschel named them Enceladus and Mimas, and found, on looking up his former observations of Saturn, that Enceladus, the exterior and brighter object, had been unmistakably seen with the twenty-foot, August 19th, 1787. Mimas is a very delicate test of instrumental perfection.

The mirror by which it was first shown measured nearly fifty inches across, and weighed 2,118 pounds. It was slung in a ring, and the sheet-iron tube in which it rested was thirty-nine and a-half feet long and four

D

feet ten inches wide.   Ladders fifty feet in length gave
access to a movable stage, from which the observer communicated through speaking tubes with his assistants.
The whole erection stood on a revolving platform ; for
the modern equatorial form of mount, by which the
diurnal course of the heavens is automatically followed,
was not then practically available, and the necessary
movements had to be imparted by hand. This
involved the attendance of two workmen, but was
otherwise less inconvenient than might be supposed,
owing to the skill with which the required mechanism
was contrived.

Herschel estimated that, with a magnifying-power
of 1,000, this grand instrument could, in the climate of
England, be effectively used during no more than one
hundred hours of every year.   A review with it of
the whole heavens would hence have occupied eight
centuries.   In point of fact, he found the opportunities
for its employment scarce.   The machine took some
time to get started, while the twenty-foot was ready in
ten minutes.   The speculum, moreover, proved un-
pleasantly liable to become dewed in moist weather, or
frozen up in cold ; and, in spite of all imaginable care,
it preserved the delicacy of its polish no more than
two years.   An economist of minutes, such as its
maker, could, then, do no otherwise than let the giant
telescope lie by unless its powers were expressly
needed.   They were surprisingly effective.   " With
the forty-foot instrument," he reported to the Royal
Society in 1800, " the appearance of Sirius announced
itself at a great distance like the dawn of the morning,
till this brilliant star at last entered the field, with all
the splendour of the rising sun, and forced me to take
my eye from that beautiful sight."   Which, however,

left the vision impaired in delicacy for nigh upon half-an-hour.

Thus the results gathered from the realisation of Herschel's crowning optical achievement fell vastly short of what his imagination had pictured. The promise of the telescope's initial disclosures was not realised in its subsequent career. Yet it was a superb instrument. The discovery with it of Mimas gave certain proof that the figure of the speculum was as perfect as its dimensions were unusual. But its then inimitable definition probably fell off later. Its "broad bright eye " was, for the last time, turned towards the heavens January 19th, 1811, when the Orion nebula showed its silvery wings to considerable advantage. But incurable dimness had already set in—incurable, because the artist's hand had no longer the strength needed to cure the growing malady. The big machine was, however, left standing, framework and all. It figured as a landmark on the Ordnance Survey Map of England; and, stamped in miniature on the seal of the Royal Astronomical Society, aptly serves to illustrate its motto, " *Quicquid nitet notandum.*" At last, on New Year's Eve, 1839, the timbers of the scaffolding being dangerously decayed, it was, with due ceremony, dismounted. A " Requiem," composed by Sir John Herschel, was sung by his family, fourteen in number, assembled within the tube, which was then riveted up and laid horizontally on three stone piers in the garden at Slough. " It looks very well in its new position," Sir John thought. Yet it has something of a *memento mori* aspect. It seems to remind one that the loftiest human aspirations are sprinkled "with the dust of death." The speculum adorns the hall of Observatory House.

D 2

Herschel married, May 8th, 1788, Mary, the only
child of Mr. James Baldwin, a merchant in the City
of London, and widow of Mr. John Pitt. She was
thirty-eight and he fifty. Her jointure relieved him
from pecuniary care, and her sweetness of disposition
secured his domestic happiness. They set up a
curious double establishment, taking a house at
Upton, while retaining that at Slough. Two
maidservants were kept in each, and a footman
maintained the communications. So at least runs
Mrs. Papendick's gossip. Miss Burney records in her
Diary a tea at Mr. De Luc's, where Dr. Herschel
accompanied a pair of vocalists "very sweetly on the
violin. His newly-married wife was with him, and
his sister. His wife seems good-natured; she was rich,
too! And astronomers are as able as other men to
discern that gold can glitter as well as stars."

He was now at the height of prosperity and
renown. Diplomas innumerable were showered upon
him by Academies and learned societies. In a
letter to Benjamin Franklin, he returned thanks for
his election as a member of the American Philosophical Society, and acquainted him with his recent
detection of a pair of attendants on the "Georgian
planet." A similar acknowledgment was addressed to
the Princess Daschkoff, Directress of the Petersburg
Academy of Sciences. The King of Poland sent him
his portrait; the Empress Catherine II. opened
negotiations for the purchase of some of his specula.
Lucien Bonaparte repaired to Slough incognito;
Joseph Haydn snatched a day from the turmoil of
his London engagements to visit the musician-
astronomer, and gaze at his monster telescopes. By
universal agreement, Dr. Burney declared, Herschel

was " one of the most pleasing and well-bred natural
characters of the day, as well as the greatest
astronomer." They had much in common, according
to Dr. Burney's daughter. Both possessed an uncommon " suavity of disposition " ; both loved music;
and Dr. Burney had a " passionate inclination for
astronomy." They became friends through the medium
of Dr. Burney's versified history of that science. In
September, 1797, he called at Slough with the
manuscript in his valise. " The good soul was at
dinner," he relates; and, to his surprise, since he was
ignorant of Herschel's marriage, the company included
several ladies, besides " a little boy." He was, nothing
loth, compelled to stay over-night ; discussed with his
host the plan of his work, and read to him its eighth
chapter. Herschel listened with interest, and modestly
owned to having learnt much from what he had
heard ; but presently dismayed the author by con-
fessing his " aversion to poetry," which he had
generally regarded as " an arrangement of fine words
without any adherence to the truth." He added,
however, that " when truth and science were united
to those fine words," they no longer displeased him.
The readings continued at intervals, alternately at
Slough and Chelsea, to the immense gratification of
the copious versifier, who occasionally allowed his
pleasure to overflow in his correspondence.

" Well, but Herschel has been in town," he wrote
from Chelsea College, December 10th, 1798, " for short
spurts and back again, two or three times, and I have
had him here two whole days. I read to him the first
five books without any one objection." And again ;
" He came, and his good wife accompanied him, and I
read four and a-half books ; and on parting, still more

humble than before, or still more amiable, he thanked
me for the instruction and entertainment I had given
him. What say you to that? Can anything be
grander?"

In spite of his "aversion," Herschel had once, and
once only, wooed the coy muse himself. The first
evening paper that appeared in England, May 3rd,
1788, contained some introductory quatrains by him.
An excuse for this unwonted outburst may be found
in the circumstance that the sheet in which they were
printed bore the name of *The Star*. They began
with the interrogation:

> "What Star art thou, about to gleam
>   In Novelty's bright hemisphere?"

and continued:

> "A Planet wilt thou roll sublime,
>   Spreading like Mercury thy rays?
> Or chronicle the lapse of Time,
>   Wrapped in a Comet's threatening blaze?"

That they are of the schoolboy order need surprise
no one. Such a mere sip at the "Pierian spring"
could scarcely bring inspiration.

Herschel's grand survey of the heavens closed with
his fourth review. His telescopic studies thereupon
became specialised. The sun, the planets and their
satellites, the lately discovered asteroids, certain double
stars, and an occasional comet, in turn received
attention. Laboratory experiments were also carried
on, and discussions of profound importance were laid
before the Royal Society. All this cost him but little
effort. The high tension of his earlier life was somewhat relaxed; he allowed himself intervals of rest,
and indulged in social and musical recreations.

Concerts were now frequently given at his house ; and the face of beaming delight with which he presided over them is still traditionally remembered. Visits to Sir William Watson at Dawlish gave him opportunities, otherwise rare, for talks on metaphysical subjects; and he stayed with James Watt at Heathfield in 1810. He had been a witness on his side in an action for infringement of patent in 1793.

Herschel rented a house on Sion Hill, Bath, for some months of the year 1799 ; and from time to time stayed with friends in London, or sought change of air at Tunbridge Wells, Brighton, or Ramsgate. In July, 1801, he went to Paris with his wife and son, made acquaintance with Laplace, and had an interview with the First Consul. It was currently reported that Bonaparte had astonished him by the extent of his astronomical learning ; but the contrary was the truth. He had tried to be impressive, but failed. Herschel gave an account of what passed to the poet Campbell, whom he met at Brighton in 1813.

" The First Consul," he said, "did surprise me by his quickness and versatility on all subjects ; but in science he seemed to know little more than any well-educated gentleman ; and of astronomy much less, for instance, than our own king. His general air was something like affecting to know more than he did know." Herschel's election in 1802 as one of the eight foreign Associates of the French Institute was probably connected with his Parisian experiences.

He inspired Campbell with the most lively enthusiasm. " His simplicity," he wrote, " his kindness, his anecdotes, his readiness to explain—and make perfectly conspicuous too—his own sublime conceptions of the universe, are indescribably charming. He is seventy-six, but fresh and stout; and there he sat, nearest the door at his friend's house, alternately smiling at a joke, or contentedly sitting without share or notice in the conversation. Any train of conversation he follows implicitly; anything you ask, he labours with a sort of boyish earnestness to explain. Speaking of himself, he said, with a modesty of manner which quite overcame me, when taken together with the greatness of the assertion, 'I have looked further into space than ever human being did before me; I have observed stars, of which the light, it can be proved, must take two millions of years to reach this earth.' I really and unfeignedly felt at the moment as if I had been conversing with a supernatural intelligence. 'Nay, more,' said he, 'if those distant bodies had ceased to exist two millions of years ago we should still see them, as the light would travel after the body was gone.' These were Herschel's words; and if you had heard him speak them, you would not think he was apt to tell more than the truth."

The appearance of a bright comet, in October, 1806, drew much company to Slough. On the 4th, Miss Herschel narrates, "Two parties from the Castle came to see it, and during the whole month my brother had not an evening to himself. As he was then in the midst of polishing the forty-foot mirror, rest became absolutely necessary after a day spent in that most laborious work; and it has ever been my opinion that on the 14th of October his nerves received a shock from which he never got the better afterwards; for on that day he had hardly dismissed his troop of men when visitors assembled, and from the time it was dark, till past midnight, he was on the grass-plot surrounded by between fifty and sixty persons, without having had time to put on proper clothing, or for the least nourishment to pass his lips. Among the company I remember were the Duke of Sussex, Prince Galitzin, Lord Darnley, a number of officers, Admiral Boston, and some ladies."

A dangerous attack of illness in the spring of 1807 left Herschel's strength permanently impaired. But he travelled to Scotland in the summer of 1810, and received the freedom of the City of Glasgow. Then, in 1814, he made a final, but fruitless attempt, to renovate the four-foot speculum. In the midst of the confusion attending upon the process, word was given to prepare for the reception of the Czar Alexander, the Duchess of Oldenburg, and sundry other grandees just then collected at Windsor for the Ascot races. The setting to rights was no small job; " but we might have saved ourselves the trouble," his sister remarks drily, "for they were sufficiently harassed with public sights and festivities."

On April 5th, 1816, Herschel was created a Knight of the Royal Hanoverian Guelphic Order, and duly attended one of the Prince Regent's levées in May. He went to town in 1819 to have his portrait painted by Artaud. The resulting fine likeness is in the possession of his grandson, Sir William James Herschel. The Astronomical Society chose him as its first President in 1821; and he contributed to the first volume of its memoirs a supplementary list of 145 double stars. The wonderful series of his communications to the Royal Society closed when he was in his eightieth year, with the presentation, June 11th, 1818, of a paper on the Relative Distances of Star-clusters. On June 1st, 1821, he inserted into the tube with thin and trembling hands the mirror of the twenty-foot telescope, and took his final look at the heavens. All his old instincts were still alive, only the bodily power to carry out their behests was gone. An unparalleled career of achievement left him unsatisfied with what he had done. Old age brought him no Sabbath rest, but only an enforced and wearisome cessation from activity. His inability to re-polish the four-foot speculum was the doom of his *chef d'œuvre.* He could not reconcile himself to it. His sunny spirits gave way. The old happy and buoyant temperament became overcast with despondency. His strong nerves were at last shattered.

On August 15th, 1822, Miss Herschel relates :— " I hastened to the spot where I was wont to find him with the newspaper I was to read to him. But I was informed my brother had been obliged to return to his room, whither I flew immediately. Lady Herschel and the housekeeper were with him, administering everything which could be thought of for supporting him. I found him much irritated at not being able to grant Mr. Bulman's* request for some token of remembrance for his father. As soon as he saw me, I was sent to the library to fetch one of his last papers and a plate of the forty-foot telescope. But for the universe I could not have looked twice at what I had snatched from the shelf, and when he faintly asked if the breaking-up of the Milky Way was in it, I said ' Yes,' and he looked content. I cannot help remembering this circumstance ; it was the last time I was sent to the library on such an occasion. That the anxious care for his papers and workrooms never ended but with his life, was proved by

* The grandson of one of Herschel's earliest English friends.

his whispered inquiries if they were locked, and the key safe."

He died ten days later, August 25th, 1822. Above his grave, in the church of Saint Laurence at Upton, the words are graven:—" Coelorum perrupit claustra " —He broke through the barriers of the skies.

William Herschel was endowed by nature with an almost faultless character. He had the fervour, without the irritability of genius; he was generous, genial, sincere; tolerant of ignorance; patient under the acute distress, to which his situation rendered him peculiarly liable, of unseasonable interruptions at critical moments: he was warm-hearted and open-handed. His change of country and condition, his absorption in science, the homage paid to him, never led him to forget the claims of kindred. Time and money were alike lavished in the relief of family necessities. He supported his brother Alexander after his retirement from the concert-stage in 1816, until his death at Hanover, March 15th, 1821. Dietrich's recurring misfortunes met his unfailing pity and help. He bequeathed to him a sum of £2,000, and to his devoted sister, Caroline, an annuity of £100.

His correspondents, abroad and at home, were numerous; nor did he disdain to remove the perplexities of amateurs. In a letter, dated January 6th, 1794, we find him explaining to Mr. J. Miller of Lincoln's Inn, "the circumstances which attend the motion of a race-horse upon a circle of longitude." And he wrote shortly afterwards to Mr. Smith of Tewkesbury:—"You find fault with the principles of gravitation and projection because they will not account for the rotation of the planets upon their axes. You might certainly with as much reason find fault with your shoes because they will not likewise serve your hands as gloves. But, in my opinion, the projectile motion once admitted, sufficiently explains the rotatory motion; for it is hardly possible mechanically to impress the one without giving the other at the same time."

On religious topics he was usually reticent; but a hint of the reverent spirit in which his researches were conducted may be gathered from a sentence in the same letter. " It is certainly," he said, " a very laudable thing to receive instruction from the great workmaster of nature, and for that reason all experimental philosophy is instituted."

To investigate was then, in his view, to " receive instruction "; and one of the secrets of his wonderful success lay in the docility with which he came to be taught.

CHAPTER III.

THE EXPLORER OF THE HEAVENS.

" A KNOWLEDGE of the construction of the heavens,"
Herschel wrote in 1811, " has always been the ultimate
object of my observations." The " Construction of
the Heavens "! A phrase of profound and novel
import, for the invention of which he was ridiculed
by Brougham in the *Edinburgh Review*; yet ex-
pressing, as it had never been expressed before, the
essential idea of sidereal astronomy. Speculation
there had been as to the manner in which the stars
were grouped together ; but the touchstone of reality
had yet to be applied to them. This unattempted,
and all but impossible enterprise Herschel deliberately
undertook. It presented itself spontaneously to his
mind·as worth the expenditure of a life's labour ; and
he spared nothing in the disbursement. The hope
of its accomplishment inspired his early exertions,
carried him through innumerable difficulties, lent
him audacity, fortified him in perseverance. For this,

" He left behind the painted buoy
  That tosses at the harbour's mouth,"

and burst his way into an unnavigated ocean.

Herschel has had very few equals in his strength
of controlled imagination. He held the balance, even
to a nicety, between the real and the ideal. Meditation
served in him to prescribe and guide experience ;
experience to ripen the fruit of meditation.

" We ought," he wrote in 1785, " to avoid two opposite extremes. If we indulge a fanciful imagination, and build worlds of our own, we must not wonder at our going wide from the path of truth and nature. On the other hand, if we add observation to observation without attempting to draw, not only certain conclusions, but also conjectural views from them, we offend against the very end for which only observations ought to be made."

This was consistently his method. If thought outran sight, he laboured earnestly that it should be overtaken by it : while sight, in turn, often took the initiative, and suggested thought. He was much more than a simple explorer. " Even at the telescope," Professor Holden says, " his object was not discovery merely, but to know the inner constitution of the heavens." He divined, at the same time that he observed.

The antique conception of the heavens as a hollow sphere upon which the celestial bodies are seen projected, survived then, and survives now, as a convenient fiction for practical purposes. But in the eighteenth century the fiction assumed to the great majority a sort of quasi-reality. Herschel made an exception in being vividly impressed with the *depth* of space. How to sound that depth was the first problem that he attacked. As a preliminary to further operations, he sought to fix a unit of sidereal measurement. The distances from the earth to the stars were then altogether unknown. All that had been ascertained was that they must be very great. Instrumental refinements had not, in fact, been carried far enough to make the inquiry profitable. Herschel did not underrate its difficulty.

He recognised that, in pursuing it, *one hundredth of a second of arc* " became a quantity to be considered." Justly arguing, however, that previous experiments on stellar parallax had been unsatisfactory and indecisive, he determined to try again.

He chose the "double star method," invented by Galileo, but never, so far, effectually put to trial. The principle of it is perfectly simple, depending upon the perspective shifting to a spectator in motion, of objects at different distances from him. In order to apprehend it, one need only walk up and down before a lamp placed in the middle of a room, watching its apparent change of position relative to another lamp at the end of the same room. Just in the same way, a star observed from opposite sides of the earth's orbit is sometimes found to alter its situation very slightly by comparison with another star close to it in the sky, but indefinitely remote from it in space. Half the small oscillation thus executed is called that star's "annual parallax." It represents the minute angle under which the radius of the terrestrial orbit would appear at the star's actual distance. So vast, however, is the scale of the universe, that this tell-tale swing to and fro is, for the most part, imperceptible even with modern appliances, and was entirely inaccessible to Herschel's observations. Yet they did not remain barren of results.

" As soon as I was fully convinced," he wrote in 1781, " that in the investigation of parallax the method of double stars would have many advantages above any other, it became necessary to look out for proper stars. This introduced a new series of observations. I resolved to examine every star in the heavens with the utmost attention that I might fix my observations upon those that would best answer my end. The subject promises so rich a harvest that I cannot help inviting every lover of astronomy to join with me in observations that must inevitably lead to new discoveries. I took some pains to find out what double stars had been recorded by astronomers; but my situation permitted me not to consult extensive libraries, nor, indeed, was it very material; for as I intended to view the heavens myself, Nature, that great volume, appeared to me to contain the best catalogue."

On January 10th, 1782, he presented to the Royal Society a catalogue of 269 double stars, of which 227 were of his own finding; and a second list of 434 followed in December, 1784. All were arranged in six classes, according to the distance apart of their components, ranging from one up to 120 seconds. The close couples he regarded as especially adapted for parallax-determinations; the wider ones might serve for criteria of stellar proper movements, or even of the sun's transport through space. For the purpose of measuring the directions in which their members lay towards each other—technically called "position-angles"—and the intervals separating them, he invented two kinds of micrometers, and notes were added as to their relative brightness and colours. He was the first to observe the lovely contrasted or harmonised tints displayed by some of these objects.

Herschel's double stars actually fulfilled none of the functions assigned to them. He was thus left without any definite unit of measurement for sidereal space; and he never succeeded in supplying the want. In 1814 he was "still engaged," though vainly, "in ascertaining a scale whereby the extent of the universe, so far as it is possible for us to penetrate into space, may be fathomed." He knew only that the distances of the stars nearest the earth could not be less, and might be a great deal more, than light-waves, propagated at the rate of 186,300 miles a second, would traverse in three or four years. Only the *manner* of stellar arrangement, then, remained open to his zealous investigations.

The initial question presenting itself to an intelligent spectator of the nocturnal sky is: What relation does the dim galactic star-stream bear to the constellations amidst which it flows? And this question our interior position makes very difficult to be answered. We see the starry universe, it has been well said, " not in *plan* but in *section.*" The problem is, from that section to determine the plan—to view the whole mentally as it would show visually from the outside. The general appearance to ourselves of the Milky Way leaves it uncertain whether it represents the projection upon the heavens of an immense stratum of equally scattered stars, or a ring-like accumulation, towards the middle of which our sun is situated. Herschel gave his preference, to begin with, to the former hypothesis, and then, with astonishing boldness and ingenuity, attempted to put it experimentally to the proof.

His method of "star-gauging" was described in 1784. It consisted in counting the stars visible in successive fields of his twenty-foot telescope, and computing the corresponding depths of space. Admitting an average regularity of distribution, this was easily done. If the stars did not really lie closer together in one region than in another, then the more

E

of them there were to be seen along a given line
of vision, the further the system could be inferred
to extend in that particular direction. The ratio
of its extension would also be given. It would vary
with the cube-roots of the number of stars in each
count.

Guided by this principle, Herschel ventured to lay
down the boundaries of the stellar aggregation to
which our sun belongs. So far as he " had yet gone
round it," in 1785, he perceived it to be " everywhere
terminated, and in most places very narrowly too.''
The differences, however, between his enumerations in
various portions of the sky were enormous. In the
Milky Way zone the stars presented themselves in
shoals. He met fields—of just one quarter the area
of the moon—containing nearly 600; so that, in
fifteen minutes, 116,000 were estimated to have
marched past his stationary telescope. Here, the
calculated "length of his sounding-line" was nearly
500 times the distance of Sirius, his standard star.
Towards the galactic poles, on the contrary, stars were
comparatively scarce ; and the transparent blackness
of the sky showed that in those quarters the supply
of stars was completely exhausted. At right angles
to the Milky Way, then, the stellar system might be
termed shallow, while in its plane, it stretched out on
all sides to an inconceivable, though not to an
illimitable extent. Its shape appeared, accordingly,
to be that of a flat disc, of very irregular contour, and
with a deep cleft matching the bifid section of the
Milky Way between Cygnus and Scorpio.

Herschel regarded this conclusion only "as an
example to illustrate the method." Yet it was derived
from the reckoning-up of 90,000 stars in 2,400 tele-

scopic fields! Its validity rested on the assumption
that stellar crowding indicated, not more stars in a
given space, but more space stocked in the same proportion with stars. But his hope of thus getting a
true mean result collapsed under the weight of his
own observations. "It would not be difficult," he
stated in 1785, "to point out two or three hundred
gathering clusters in our system." The action of a
"clustering power" drawing its component stars "into
many separate allotments" grew continually clearer
to him, and he admitted unreservedly in 1802 that
the Milky Way "consists of stars very differently
scattered from those immediately about us."

In 1811, he expressly abandoned his original
hypothesis. "I must freely confess," he wrote, "that
by continuing my sweeps of the heavens my opinion
of the arrangement of the stars has undergone a
gradual change. An equal scattering of the stars may
be admitted in certain calculations; but when we
examine the Milky Way, or closely compressed
clusters, it must be given up."

And in 1817: "Gauges, which on a supposition of
an equality of scattering, were looked upon as gauges
of distance, relate, in fact, more immediately to the
scattering of the stars, of which they give valuable
information."

The "disc-theory" was then virtually withdrawn
not many years after it had been propounded. "The
subtlety of nature," according to Bacon's aphorism,
"transcends the subtlety both of the intellect and
of the senses." Herschel very soon perceived the
inadequacy of his colossal experiment; and he tranquilly acquiesced, not being among those who seek to
entrench theory against evidence. He found that he

E 2

had undervalued the complexity of the problem. Yet it remained before his mind to the end. The supreme object of his scientific life was to ascertain the laws of stellar distribution in cubical space, and he devoted to the subject the two concluding memoirs of the sixty-nine contributed by him to the "Philosophical Transactions." He was in his eightieth year when he opened, with youthful freshness, a new phase of arduous investigation.

"The construction of the heavens," he wrote in June, 1817, "can only be known when we have the situation of each body defined by its three dimensions. Of these three, the ordinary catalogues give but two, leaving the distance or profundity undetermined." This element of "profundity" he went on to determine by the absolutely novel method of what may be called "photometric enumeration."

He began by asserting what is self-evident—that faint stars are, "one with another," more remote than bright ones ; and he argued thence, reasonably enough, that the relative mean distances of the stars, taken order by order, might be inferred from their relative mean magnitudes. Next he pointed out that more space would be available for their accommodation in proportion to the cubes of their mean distances. Here lies the value of the method. It sets up, as Herschel said, "a standard of reference" with regard to stellar distribution. It makes it possible to compare actual stellar density, at a given mean distance, with a "certain properly modified equality of scattering." By patiently calling over the roll of successive magnitudes, information may be obtained regarding over- and under-populated districts of space.

Herschel's reasonings on the subject are perfectly

valid, but for practical purposes far in advance of the time. Their application demanded a knowledge of stellar light-gradations, which, even now, has been only partially attained. His surprising anticipation of this mode of inquiry came, therefore, to nothing.

His device of "limiting apertures" was a simultaneous invention. It was designed as a measure of relative star-distances. Pointing two similar telescopes upon two unequal stars, he equalised them to the eye by stopping down the aperture of the instrument directed towards the brighter object. Assuming each to emit the same quantity of light, their respective distances would then be inversely as the diameters of the reflecting surfaces by which they were brought to the same level of apparent lustre. But the enormous real diversities of stellar size and brightness render this plan of action wholly illusory. Even for average estimates, proper motion is apparently a safer criterion of distance than magnitude.

Herschel employed the method of apertures with better success to ascertain the comparative extent of natural and telescopic vision. The boundary of the former was placed at "the twelfth order of distance." Sirius, that is to say, removed to twelve times its actual remoteness, would be a barely discernible object to the naked eye. The same star carried seventy-five times further away still, could be seen as a faint light-speck with his twenty-foot telescope ; and, transported 192 times beyond the visual limit, would make a similar appearance in the field of the forty-foot. These figures, multiplied by twelve, represented, in his expressive phrase, the " space-penetrating power" of his instruments. Their range extended respectively to 900 and 2,300 times the distance of his " standard star." He estimated, moreover, that, through the agency of the larger, light might become sensible to the eye after a journey lasting nearly seven thousand years ! So that, as he said, his telescopes penetrated both time and space.

His last observation of the Milky Way showed it to be in parts "fathomless," even with the forty-foot. No sky-background could be seen, but only the dim glow of "star-dust." This effect he attributed to the immeasurable extension, in those directions, of the stellar system. The serried orbs composing it, as they lay further and further from the eye, became at last separately indistinguishable. Herschel, as has been said, formulated no second theory of galactic structure after that of 1784-5 had been given up. What he thought on the subject, with ripened experience for his guide, can only be gathered piecemeal from his various writings. The general appearance of

> That " broad and ample road, whose dust is gold,
> And pavement stars,"

he described as " that of a zone surrounding our situation in the solar system, in the shape of a succession of differently condensed patches of brightness, intermixed with others of a fainter tinge." And he evidently considered this *seeming* to be in fair accord with reality. The "patches of brightness" stood for genuine clusters, incipient, visibly forming, or formed. They are made up of stars not less lustrous, but much more closely collected than Sirius, Arcturus, or Capella. The smallness of galactic stars would thus be an effect of distance, while their crowding is a physical fact. The whole of these clusters are (on Herschel's view) aggregated into an irregular, branching ring, distinct from, although bound together into one system with the brilliants of the constellations. "Our sun," he emphatically affirmed in 1817, "with all the stars we can see with the eye, are deeply immersed in the Milky Way, and form a component part of it."

He took leave of the subject which had engrossed so many of his thoughts in a paper read before the Royal Society, June 11th, 1818. In it he showed how the "equalising" principle could be applied to determine the relative distances of "globular and other clusters," provided only that their component stars are of the rank of Sirius. It is improbable, however, that this condition is fulfilled. In open groups, such as the Pleiades, enormous suns are most likely connected with minute self-luminous bodies; but the stars compressed into "globular clusters" appear to be more uniform, and may, perhaps, be intermediate in magnitude. Yet here again, the only thing certain is the prevalence of endless variety. Celestial systems are not turned out by the dozen, like articles from a factory. Each differs from the rest in scale, in structure, in mechanism. Attempts to reduce all to any common standard must then prove futile. Disparities of distance are of course concerned in producing their varieties of aspect, coarse-looking "balls of stars" being, necessarily, on the whole, less remote than those of smoother texture. Finer graining, however, may also be due to a composition out of smaller and closer masses. The two causes concur, and the share of each in producing a certain effect cannot, in any individual case, be apportioned.

Herschel was indeed far too philosophical to adopt rigid lines of argument. His reasoning did not extend

"so far as to exclude a real difference, not only in the
size, but also in the number and arrangement of the
stars in different clusters." Nevertheless, the discussion founded upon it is no longer convincing. To
modern astronomers it appears to travel quite wide of
the mark. Its interest consists in the proof given by
it that the problem of sidereal distances, the original
incentive to Herschel's reviews of the heavens,
attracted his attention to the very end of his thinking
life. Throughout his long career, the profundities of
the universe haunted him, He sought, *per fas, per
nefas*, trustworthy measures of the "third dimension"
of celestial space. The object of his search was out of
reach, and has not even now been fully attained ; but
the path it led him by was strewn with discoveries.

The nets spread in his "sweeps" brought in,
besides double stars, plentiful takes of the filmy
objects called "nebulæ." He recognised with amazement their profusion in certain tracts of the sky ;
increased telescopic light-grasp never failed to render
a further supply visible ; the heavens teemed with
them. He presented a catalogue of 1,000 to the
Royal Society in 1786, a second equally compre-
hensive in 1789, and a supplementary list of 500 in
1802. Their natural history fascinated him. What
they were, what they had been, and what they should
come to, formed the subject of many of those ardent
meditations which supplied motive power for his
researches. He not only laid the foundation of
nebular science, but carried the edifice to a considerable height, distinguishing the varieties of its objects,
and classifying them according to their gradations
of brightness. Some presented a most fantastic
appearance.

"I have seen," he wrote in 1784, "double and treble nebulæ variously arranged ; large ones with small, seeming attendants ; narrow, but much extended lucid nebulæ or bright dashes ; some of the shape of a fan, resembling an electric brush, issuing from a lucid point ; others of the cometic shape, with a seeming nucleus in the centre, or like cloudy stars surrounded with a nebulous atmosphere ; a different sort, again, contained a nebulosity of the milky kind, like that wonderful, inexplicable phenomenon about Theta Orionis ; while others shine with a fainter mottled kind of light which denotes their being resolvable into stars."

He, " through the mystic dome," discerned

" Regions of lucid matter taking form,
    Brushes of fire, hazy gleams,
Clusters and beds of worlds, and bee-like swarms
    Of suns and starry streams."

Annular and planetary nebulæ were *as such*, first described by him. "Among the curiosities of the heavens," he announced in 1785, "should be placed a nebula that has a regular concentric dark spot in the middle, and is probably a ring of stars." This was the famous annular nebula in Lyra, then a unique specimen, now the type of a class.

The planetary kind, so-called from their planet-like discs, were always more or less of an enigma to him. The vividness and uniformity of their light appeared to cut them off from true nebulæ ; on mature consideration, he felt driven to suppose them "compressed star-groups." "If it were not, perhaps, too hazardous," he went on, " to pursue a former sur-mise of a renewal in what I figuratively called the laboratories of the universe, the stars forming these extraordinary nebulæ, by some decay or waste of nature, being no longer fit for their former purposes, and having their projectile forces, if any such they had, retarded in each other's atmospheres, rush at last together, and either in succession, or by one general tremendous shock, unite into a new body. Perhaps the extraordinary and sudden blaze of a new star in Cassiopeia's Chair, in 1572, might possibly be of such a nature."

At that early stage of his inquiries, Herschel regarded all nebulæ indiscriminately as composed of genuine stars. It was almost inevitable that he should do so. For each gain in telescopic power had the effect of transferring no insignificant proportion of them from the nebular to the stellar order. There was no apparent reason for drawing a line anywhere. The inference seemed irresistible, that resolvability was simply a question of optical improvement. As Messier's *nébuleuses sans étoiles* had yielded to Herschel's telescopes, so—it might fairly be anticipated—the "milky" streaks and patches seen by Herschel would curdle into stars under the compulsion of the still mightier instruments of the future. He was led on—to use his own expressions in 1791—" by almost imperceptible degrees from evident clusters, such as the Pleiades, to spots without a trace of stellar formation, the gradations being so well connected as to leave no doubt that all these phenomena were equally stellar." They were what Lambert and Kant had supposed them to be—island-universes, vast congeries of suns, independently organised, and of galactic rank. They were, each and all, glorious systems, barely escaping total submergence in the illimitable ocean of space. Under the influence of these grandiose ideas, Herschel told Miss Burney, in 1786, that with his "large twenty-foot" he had "discovered 1,500 universes!" Fifteen hundred "whole sidereal systems, some of which might well outvie our Milky Way in grandeur."

His contemplations of the heavens showed him everywhere traces of progress—of progress rising towards perfection, then sinking into decay, though with a sure prospect of renovation. He was thus led to arrange the nebulæ in a presumed order of development. The signs of interior condensation traceable in nearly all, he attributed to the persistent action of central forces. Condensation, then, gave evidence of age. Aggregated stars drew closer and closer together with time. So that scattered or branching formations were to be regarded as at an early stage of systemic existence; globular clusters, as representing universes still in the prime of life; while objects of the planetary kind were set down as "very aged, and drawing on towards a period of change, or dissolution."

Our own nebula he characterised as "a very extensive, branching congeries of many millions of stars," bearing upon it "fewer marks of profound antiquity than the rest." Yet, in certain regions, he found "reason to believe that the stars are now drawing towards various secondary centres, and will in time separate into different clusters." As an example of the ravages of time upon the galactic structure, he adverted to a black opening, four degrees wide, in the Zodiacal Scorpion, bordered on the west by an exceedingly compact cluster (Messier's No. 80), possibly formed, he thought, of stars drawn from the adjacent vacancy. The chasm was to him one of the

most impressive of celestial phenomena. His sister
preserved an indelible recollection of hearing him,
in the course of his observations, after a long,
awful silence, exclaim, "Hier ist wahrhaftig ein
Loch im Himmel!" (Here, truly, is a hole in the
sky); and he recurred to its examination night after
night and year after year, without ever clearing up,
to his complete satisfaction, the mystery of its
origin. The cluster significantly located at its edge
was lit up in 1860 by the outburst of a temporary
star.

This was not the sole instance noted by Herschel
of the conjunction of a chasm with a cluster; and
chasms and clusters alike told the same story of
dilapidation. He foresaw, accordingly, as inevitable,
the eventual "breaking-up" of the Milky Way into
many small, but independent nebulæ. "The state
into which the incessant action of the clustering power
has brought it at present," he wrote in 1814, "is a
kind of chronometer that may be used to measure
the time of its past and future existence; and although
we do not know the rate of going of this mysterious
chronometer, it is, nevertheless, certain that since
the breaking up of the Milky Way affords a proof
that it cannot last for ever, it equally bears witness
that its past duration cannot be admitted to be
infinite."

Thus the idea of estimating the relative "ages"
of celestial objects—of arranging them according
to their progress in development, originated with
Herschel in 1789. "This method of viewing the
heavens," he added, "seems to throw them into
a new kind of light. They are now seen to resemble
a luxuriant garden which contains the greatest variety

of productions in different flourishing beds; and one advantage we may at least reap from it is that we can, as it were, extend the range of our experience to an immense duration. For, is it not almost the same thing whether we live successively to witness the germination, blooming, foliage, fecundity, fading, withering, and corruption of a plant, or whether a vast number of specimens, selected from every stage through which the plant passes in the course of its existence, be brought at once to our view?"

But while he followed the line of continuity thus vividly traced, another crossing, and more or less interfering with it, opened out before him. The discovery of a star in Taurus, "surrounded with a faintly luminous atmosphere," led him, in 1791, to revise his previous opinions regarding the nature of nebulæ. He was not at all ashamed of this fresh start. No fear of "committing himself" deterred him from imparting the thoughts that accompanied his multudinous observations. He felt committed to nothing but truth. He was advancing into an untrodden country. At every step he came upon unexpected points of view. The bugbear of inconsistency could not prevent him from taking advantage of each in turn to gain a wider prospect.

Until 1791 Herschel never doubted that gradations of distance fully accounted for gradations of nebular resolvability. He had been led on, he explained, by almost imperceptible degrees from evident clusters to spots without a trace of stellar formation, no break anywhere suggesting tho possibility of a radical difference of constitution. "When I pursued these researches," he went on, "I was in the situation of a natural philosopher who follows the various species of animals and insects from the height of their perfection down to the lowest ebb of life; when, arriving at the vegetable kingdom, he can scarcely point out to us the precise boundary where the animal ceases and the plant begins; and may even go so far as to suspect them not to be essentially different. But, recollecting himself, he compares, for instance, one of the human species to a tree, and all doubt upon the subject vanishes. In the same manner we pass by gentle steps from a coarse cluster to an object such as the nebula in Orion, where we are still inclined to remain in the once adopted idea of stars exceedingly remote and inconceivably crowded, as being the occasion of that remarkable appearance. It seems, therefore, to require a more dissimilar object to set us right again. A glance like that of the naturalist, who casts his eye from the perfect animal to the perfect vegetable, is wanting to remove the veil from the mind of the astronomer. The object I have mentioned above is the phenomenon that was wanting. View, for instance, the nineteenth cluster of my sixth class, and afterwards cast your eye on this cloudy star, and the result will be no less decisive than that of the naturalist. Our judgment, I venture to say, will be that *the nebulosity about the star is not of a starry nature.*"

In this manner he inferred the existence of real nebulous matter—of a " shining fluid " of unknown and unimaginable properties. Was it perhaps, he asked himself, a display of electrical illumination, like the aurora borealis, or did it rather resemble the " magnificent cone of the zodiacal light ? " A boundless field of speculation was thrown open. " These nebulous stars," he added, "may serve as a clue to unravel other mysterious phenomena."

As their close allies, he now recognised planetary nebulæ, the "milkiness, or soft tint of their light," agreeing much better with the supposition of a fluid, than of a stellar condition. And he rightly placed in the same category the Orion nebula, and certain "diffused nebulosities" which he had observed just to tarnish the sky over wide areas. These last might, he considered, be quite near the earth, and the object in Orion not more distant than perhaps an average second magnitude star.

The relations of the sidereal to the nebular "principle" exercised Herschel's thoughts during many years. He had no sooner reasoned out the existence in inter-stellar space of a rarefied, self-luminous substance, than he began to interrogate himself as to its probable function. Nature was to him the expression of Supreme Reason. He could only conceive of her doings as directed towards an intelligible end. Hence his confidence that rational investigation must lead to truth.

Already in 1791 he hinted at the conclusion which he foresaw. The envelope of a "cloudy star" was, he declared, "more fit to produce a star by its condensation than to depend upon the star for its existence." And the surmise was confirmed by his detection, in a planetary nebula, of a sharp nucleus, or "generating star," possibly to be completed in time by the further accumulation of luminous matter.

His conjectures developed in 1811 into a formal theory. The cosmical fluid was met with in all stages of condensation. Nebulous tracts of almost evanescent

lustre were connected in an unbroken series with
slightly "burred" objects, wanting only a few last
touches to make them finished stars. The extremes,
as he said, had been, by his "critical examination of
the nebulous system," "connected by such nearly
allied intermediate steps, as will make it highly
probable that every succeeding state of the nebulous
matter is the result of the action of gravitation upon
it while in a foregoing one."

In 1814 he traced the progress towards maturity
of binary systems. Originating in double nebulæ
incompletely dissevered—Siamese-twin objects, of
which he had collected 139 examples—they next
appeared as nebulously-connected stars, finally as a
pair materially isolated, and linked together by the
sole tie of gravitation. Scattered clusters represented,
in his scheme of celestial progress, a state antecedent
to that of globular clusters. "The still remaining
irregularity of their arrangement," he said, "additionally
proves that the action of the clustering power has not
been exerted long enough to produce a more artificial
construction." He made, too, the important admission
that clusters apparently "in, or very near the Milky
Way," were truly part and parcel of that complex
agglomeration.

But what of his "fifteen hundred universes," which
had now logically ceased to exist? The stellar and
nebular "principles" had virtually coalesced; both
were included in the galactic system. The question
of "island universes" was accordingly left in abeyance;
although Herschel certainly believed in 1818 that
among the multitude of "ambiguous objects"—
we should call them irresolvable nebulæ—many exterior firmaments were included. Yet what he had

ascertained about the distribution of nebulæ should alone have sufficed to shatter this remnant of a conviction.

The fact became clear to him during the progress of his "sweeps" that nebulæ, to some extent, *replace stars*. He found them to occur in " parcels," more or less embedded with stars, "beds" and "parcels" together being surrounded by blank spaces. This arrangement grew so familiar to him that he used to notify his assistant, when stars thinned out in the zone he was traversing, " to prepare for nebulæ." A wider relationship, brought within view by the large scale of his labours, was defined by his fortunate habit of charting, for convenience of identification, each newly-discovered batch of nebulæ.

" A very remarkable circumstance," he wrote in 1784, "attending the nebulæ and clusters of stars, is that they are arranged into strata, which seem to run on to a great length; and some of them I have already been able to pursue, so as to guess pretty well at their form and direction. It is probable enough that they may surround the whole apparent sphere of the heavens, not unlike the Milky Way."

In the following year he spoke no longer of a zone, but of two vast groupings of nebulæ about the opposite poles of the Milky Way. That is to say, where stars are scarcest nebulæ are most abundant. The correspondence did not escape him; but he did not recognise its architectonic meaning. He had traced out the main plan of the stellar world; he had discovered, not merely thousands of nebulæ, but the nebular system; he had shown that stars and nebulæ were intimately associated; he had even made it clear that

F

nebular distribution was governed by the lines of galactic structure. It only remained to draw the obvious inference that these related parts made up one whole—that no more than a single universe is laid open to human contemplation. This was done by Whewell thirty years after his death.

CHAPTER IV.

HERSCHEL'S SPECIAL INVESTIGATIONS.

DOUBLE stars were, when Herschel began to pay attention to them, regarded as mere chance productions. No suspicion was entertained that a real, physical bond united their components. Only the Jesuit astronomer, Christian Mayer, maintained that bright stars were often attended by faint ones; and since his observations were not such as to inspire much confidence, his assertions counted for very little. " In my opinion," Herschel wrote in 1782, " it is much too soon to form any theories of small stars revolving round large ones." He, indeed, probably even then, suspected that close *equal* stars formed genuine couples; but he waited, if so, for evidence of the connection. The chief subject of his experiments on parallax was Epsilon Boötis, an exquisitely tinted, unequal pair. But he soon became aware that either stellar parallax was elusively small, or that he was on the wrong track for detecting it. And, since his favourite stars have proved to be a binary combination, it was, of course, drawing water in a sieve to make one the test of perspective shifting in the other.

The number of Herschel's double stars alone showed them to be integral parts of an express design. Such a crop of casualties was out of all reasonable question. And it was actually pointed out in 1784 by John Michell, a man of extraordinary sagacity,

F 2

that the odds in favour of their physical union were truly "beyond arithmetic."

Herschel meantime kept them under watch and ward, and after the lapse of a score of years found himself in a position to speak decisively. On July 1, 1802, he informed the Royal Society that "casual situations will not account for the multiplied phenomena of double stars," adding, "I shall soon communicate a series of observations proving that many of them have already changed their situation in a progressive course, denoting a periodical revolution round each other." A year later he amply fulfilled this pledge. Discussing in detail the displacements brought to light by his patient measurements, he made it clear that they could be accounted for only by supposing the six couples in question to be "real binary combinations, intimately held together by the bond of mutual attraction." His conclusion was, in each case, ratified by subsequent observation. The stars instanced by him—Castor, Gamma Leonis, Epsilon Boötis, Delta Serpentis, Gamma Virginis, and Zeta Herculis—are all noted binaries. Not satisfied with establishing the fact, Herschel assigned the periods of their revolutions. But he could only do so on the hypothesis of circular motion, while the real orbits are highly elliptical. His estimates then went necessarily wide of the mark. For one pair only, he was able to use an observation anterior to his own. Bradley had roughly fixed, in 1759, the relative position of the components of Castor, the finest double star in the northern heavens; and the preservation of the record in Dr. Maskelyne's note-book extended by twenty years the basis of Herschel's conclusions regarding this system.

He continued, in 1803, his discussions of double stars; announced a leisurely circulation of both the pairs composing the typical "double-double star," Epsilon Lyræ; and conjectured the union of the two into one grand whole—a forecast verified by the evidence of common proper motion. The Annus Magnus of the quadruple system cannot, according to Flammarion, be less than a million of years.

The discovery of binary stars was, in Arago's phrase, "one with a future." In itself an amazing revelation, it marked the beginning of a series of investigations of immense variety and importance. By it, a science of sidereal mechanics was shown to be possible; the sway of gravitation received an unlimited extension; and the perception of order, which is the precursor of knowledge, ranged at once over the whole visible creation. Herschel, it is true, had not the means of formally proving that stellar orbits are described in obedience to the Newtonian law. His affirmative assertion rested only on the analogy of the solar system. But the rightness of his judgment has never seriously been called in question.

His research into the transport of the solar system through space proved, as Bessel said, that the activity of his mind was independent of the stimulus supplied by his own observations. It was one of his most brilliant performances.

The detection of progressive star-movements was due to Halley. It was announced in 1718. The bright objects spangling the sky are then "fixed" only in name. "But if the proper motion of the stars be admitted," asked Herschel, "who can deny that of our sun?" The same idea had occurred to several earlier astronomers, but only one, Tobias Mayer, of Göttingen, had tried to test it practically; and he had failed. " To discern the proper motion of the sun between so many other motions of the stars," Herschel might well designate " an arduous task." Yet it was not on that account to be neglected. The conditions of the problem were perfectly clear to him. If the sun alone were in motion, the stars should unanimously appear to drift backward from the " apex," or point on the sphere towards which his journey was directed. The heavens would open out in front of his advance, and close up behind. The effect was compared by Mayer to the widening prospect and narrowing vista of trees to a man walking through a forest. On this supposition, the perspective displacements of any two stars suffi- ciently far apart in the sky would suffice to determine the solar apex. For it should coincide with the inter- section of the two great circles continuing the direc- tions of those displacements. But the question is far from being of this elementary nature. The stars are all flitting about on their own account, after—to our apprehension—a haphazard fashion. The sole element of general congruity traceable among them is that "systematic, or higher, parallax," by which each of them is, according to a determinate proportion, in- evitably affected. If this can be elicited, the line of the sun's progress becomes at once known.

Herschel treated the subject in the simplest possible manner. Striking a balance between the proper motions of only seven stars, he deduced, in 1783, from simple geometrical considerations, an apex for the sun's way, marked by the star Lambda Herculis. But while he seemed to proceed by rule, he was really led by the unerring instinct of genius. His mode of conducting an investigation, small in compass, yet almost inconceivably grand in import, distances praise. Its directness and apparent artlessness strike us dumb with wonder. Eminently suited to the materials at his command, it was summary, yet, within fairly narrow limits, secure. And the result has stood the test of time. It ranks, even now, as a valuable approximation to the truth. He himself regarded his essay as nothing more than an experimental effort. In a letter to Dr. Wilson, of Glasgow, he expressed his apprehensions lest his paper on the sun's motion "might be too much out of the way to deserve the notice of astronomers."

Provided with Maskelyne's table of thirty-six proper motions, he resumed the subject in 1805. He now employed a graphical method, drawing great circles to represent the observed stellar movements, and planting his apex impartially in the midst of their intersections. It was, however, less happily located than that of 1783. The constellation Hercules again just included it; but it lay certainly too far west, and probably too far north. The memoir conveying the upshot of the research is, none the less, a masterpiece. Philosophy and common-sense have rarely been so fortunately blended as in this discussion. Without any mathematical apparatus, the plan of attack upon a recondite problem is expounded with the utmost generality and precision. The reasoning is strong and sure; intelligible to the ignorant, instructive to the learned.

In his earlier paper, Herschel, while venturing only to " offer a few distant hints " as to the *rate* of the sun's travelling, expressed the opinion that it could " certainly not be less than that which the earth has in her annual orbit." That is to say, his minimum estimate was then nineteen miles per second. A direct inquiry, on the other hand, convinced him, in 1806, that the solar motion, viewed at right angles from the distance of Sirius, would cover yearly an arc of 1″. 112. This he called "its quantity;" the corresponding velocity remained undetermined. We can, however, now, since the real distance of his assumed station has been determined, translate this angular value into a linear speed of about nine miles a second. The mean of his two estimates, or fourteen miles a second, probably differs little from the actual rate at which the solar system is being borne to its unimaginable destination.

His conclusions regarding the solar translation obtained little notice, and less acceptance from his contemporaries and immediate successors. His son rejected them as untrustworthy; Bessel, the greatest authority of his time in the science of "how the heavens move," declared in 1818 that the sun's apex might be situated in any other part of the sky with as much probability as in the constellation Hercules. Not until Argelander, by a strict treatment of multiplied and improved data, arrived in 1837 at practically the same result, did Herschel's anticipatory efforts obtain the recognition they deserved. Scarcely in any department has there been put on record so well-directed a leap into the dark of coming discovery.

The systematic light-measurement of the stars began with the same untiring investigator. He described in 1796 the method since named that of "sequences," and presented to the Royal Society the first of six Photometric Catalogues embracing nearly all the 2,935 stars in Flamsteed's "British Catalogue." They gave comparative brightnesses estimated with the naked eye; classification by magnitudes was put aside as vague and misleading. The " sequences " serving for their construction were lists of stars arranged, by repeated trials, in order of lustre, and rendered mutually comparable by the inclusion in each of a few members of the preceding series. Their combination into a catalogue was then easily effected. "Simple as my method is in principle," he remarked, " it is very laborious in its progress." On a restricted scale it is still employed for following the gradations of change in variable stars.

These researches lay, as Professor Holden expresses it, "directly on the line of Herschel's main work." The separation of the stars into light-ranks intimates at once something as to their distribution in space; but the intimations may prove deceptive unless the divisions be accurately established. Hence, stellar photometry is an indispensable adjunct to the study of sidereal construction. Herschel prosecuted the subject besides with a view to ascertaining the constancy of stellar lustre. He had been struck with singular discordances between magnitudes assigned at different dates. Not to mention stars obviously variable, there were others which seemed to be affected by a slow, secular waxing or waning. In some of the instances alleged by him, the alteration was no doubt fictitious — a record of antique errors; but there was a genuine residuum. Thus, the immemorially observed constituents of the Plough preserve no fixed order of relative brilliancy, now one, now another of the septett having, at sundry epochs, assumed the primacy; while a small star in the same group, Alcor, the " rider " of the second " horse," has, in the course of a millennium, plainly thrown off some part of its former obscurity. The Arabs in the desert regarded it as a test of penetrating vision; and they were accustomed to oppose "Suhel" to "Suha" (Canopus to Alcor) as occupying respectively the highest and lowest posts in the celestial hierarchy. So that *Vidit Alcor, at non lunam plenam*, came to be a proverbial description of one keenly alive to trifles, but dull of apprehension for broad facts. Now, however, Alcor is an easy naked-eye object. One needs not be a "tailor of Breslau," or a Siberian savage, to see it. The little star is unmistakably more luminous than of old.

An inversion of brilliancy between Castor and Pollux, and between the two leading stars in the Whale, is further generally admitted to have taken place during the eighteenth century. The prevalence of such vicissitudes was deeply impressive to Herschel, especially through their bearing upon the past and future history of our own planet. "If," he said, "the similarity of stars with our sun be admitted, how necessary will it be to take notice of the fate of our neighbouring *suns*, in order to guess at that of our own. The *star* which we have dignified by the name of *Sun* may to-morrow begin to undergo a gradual decay of brightness, like Alpha Ceti, Alpha Draconis, Delta Ursæ Majoris, and many other diminishing stars. It may suddenly increase like the wonderful star in Cassiopeia, or gradually come on like Pollux, Beta Ceti, etc. And, lastly, it may turn into a periodical one of twenty-five days' duration (the solar period of rotation), as Algol is one of three days, Delta Cephei of five days, etc." He found it, accordingly, " perhaps the easiest way of accounting for past changes in climate to surmise that our sun has been formerly sometimes more, sometimes less, bright than it is at present." Herschel attempted, in 1798, to analyse star-colours by means of a prism applied to the eye-glasses of his reflector. Nothing of moment could at that time come of such experiments ; but they deserve to be remembered as a sort of premonition of future methods of research into the physical condition of the stars.

His attention to the sun might have been exclusive, so diligent was his scrutiny of its shining surface. Many of its peculiarities were first described by him, and none escaped him, except the "deeper deep," or black nucleus of spots, detected by Dawes in 1852. The dusky "pores" and brilliant "nodules," the corrugations, indentations, and ridges ; the manifold aspects of spots, or "openings ; " their "luminous shelving sides," known as penumbræ ; were all noted in detail, ranged in proper order, and studied in their mutual relations. Spots presented themselves to him as evident depressions in the luminous disc ; faculæ, "so far from resembling torches," appeared "like the shrivelled elevations upon a dried apple, extended in length, and most of them joined together, making waves, or waving lines." Towards the north and south, he went on, " I see no faculæ ; there is all over the sun a great unevenness, which has the appearance of a mixture of small points of an unequal light ; but they are evidently a roughness of high and low parts."

His theory of the solar constitution was a development of Wilson's. It was clearly conceived, firmly held, and boldly put forward. The definite picturesqueness, moreover, of the language in which it was clothed, at once laid hold of the public imagination, and gave it a place in public favour from which it was dislodged only by the irresistible assaults of spectrum analysis.

The sun was regarded by Herschel as a cool dark body surrounded by an extensive atmosphere made up of various elastic fluids. Its upper stratum—Schröter named it the "photosphere"—was of cloud-like composition, and consisted of lucid matter precipitated from the elastic medium by which it was sustained. Its depth was estimated at two or three thousand miles, and the nature of its emissions suggested a comparison with the densest coruscations of the aurora borealis. Below lay a region of "planetary," or protective clouds. Dense, opaque, and highly reflective, "they must add," he said, " a most capital support to the splendour of the sun by throwing back so great a share of the brightness coming to them." Their movements betrayed the action of vehement winds; and indeed the continual "luminous decompositions" producing the radiating shell, with the consequent regeneration of atmospheric gases beneath, "must unavoidably be attended with great agitations, such as with us might even be called hurricanes." The formation and ascent of "empyreal gas" would cause, when moderate in quantity, pores, or small openings in the brilliant layers. But should it happen to be generated in uncommon quantities, "it will burst through the planetary regions of clouds, and thus will produce great openings; then, spreading itself above them, it will occasion large shallows, and, mixing afterwards gradually with other superior gases, it will promote the increase, and assist in the maintenance of the general luminous phenomena."

The solid globe thus girt round with cloud and fire was depicted as a highly eligible place of residence.

An equable climate, romantic scenery, luxuriant vegetation, smiling landscapes, were to be found there. It might, accordingly, be admitted without hesitation that "the sun was richly stored with inhabitants." For the lucid shell visible from the exterior possessed, according to this theory, none of the all-consuming ardour now attributed to it. Its blaze was a superficial display; beneath, "the immense curtain of the planetary clouds was everywhere closely drawn " round a world perfectly accommodated to vital needs.

In order to reconcile this supposed state of things with the observed order of nature, it was suggested that traces of it subsist in the planets, "all of which, we have pretty good reason to believe, emit light in some degree." The night-side illumination of Venus, the sinister glare of the eclipsed moon, the auroral glimmerings of the earth, were adduced as evidence to this effect. The contrast between the central body and its dependants was softened down to the utmost.

"The sun, viewed in this light," Herschel wrote in 1794, " appears to be nothing else than a very eminent, large, and lucid planet, evidently the first, or, in strictness of speaking, the only primary one of our system ; all others being truly secondary to it. Its similarity to the other globes of the solar system with regard to its solidity, its atmosphere, and its diversified surface ; the rotation upon its axis, and the fall of heavy bodies, lead us on to suppose that it is also most probably inhabited, like the rest of the planets, by beings whose organs are adapted to the peculiar circumstances of that vast globe."

To us, nearing the grey dawn of the twentieth century, the idea seems extravagant ; it was, in the eighteenth, plausible and alluring. The philosophers of that age regarded the multiplicity of inhabited worlds as of axiomatic certainty. The widest possible diffusion of life followed, they held, as a corollary from the beneficence of the Creator; while their sense of economy rendered them intolerant of *wasted* globes. Herschel was then reluctant to attribute to the sun a purely *altruistic* existence. Only from the point of view of our small terrestrial egotism could so glorious a body figure as solely an attractive centre, and a focus of warmth and illumination to a group of planets. Besides, looking abroad through the universe, we see multitudes of stars which can exercise no ministerial functions. Those united to form compressed clusters, or simply joined in pairs, are unlikely, it was argued, to carry a train of satellites with them in their complex circlings. Unless, then, "we would make them mere useless brilliant points," they must " exist for themselves," and claim primary parts in the great cosmical life-drama.

Herschel's sun is to us moderns a wholly fabulous body. Still, there is a fantastic magnificence about the conception so strongly realised by his powerful imagination. Moreover, its scientific value was by no means inconsiderable. It represented the first serious effort to co-ordinate solar phenomena; it implied the spontaneous action of some sort of machinery for the production of light and heat. Spots were associated with a circulatory process; the photosphere was portrayed under its true aspect. The persistence of its hollows and heights, its pores and rugosities, convinced Herschel that the lustrous substance composing it was "neither a liquid nor an elastic fluid," which should at once subside into an unbroken level. " It exists, therefore," he inferred, " in the manner of lucid clouds swimming in the transparent atmosphere of the sun."

" The influence of this eminent body on the globe we inhabit," he wrote, continuing the subject in 1801, " is so great, and so widely diffused, that it becomes almost a duty to study the operations which are carried on upon the solar surface." This duty he fulfilled to perfection. His telescopic readings from the changeful solar disc were of extraordinary precision and comprehensiveness. They show his powers as an observer perhaps at their best. And, since reasoning was with him inseparable from seeing, the appearances he noted took, as if of their own accord, their proper places. The history of spots was completely traced. He recorded their birth by the enlargement of pores : their development and sub-division; established their connexion with faculous matter, piled up beside them like mountain-ranges round an Alpine lake, or flung across their cavities like blazing suspension-bridges ; and watched finally their closing-up and effacement, not even omitting the post-mortem examination of the disturbances they left behind.

One of Herschel's curiously original enterprises was his attempt to ascertain a possible connexion between solar and terrestrial physics. " I am now much inclined to believe," he stated in 1801, " that openings with great shallows, ridges, nodules, and corrugations, may lead us to expect a copious emission of heat, and, therefore, mild seasons. And that, on the contrary, pores, small indentations, and a poor appearance of the luminous clouds, the absence of ridges and nodules, and of large openings and shallows, will denote a spare emission of heat, and may induce us to expect severe seasons. A constant observation of the sun with this view, and a proper information respecting the general mildness or severity of the seasons in all parts of the world, may bring this theory to perfection, or refute it, if it be not well founded."

But the available data regarding weather-changes turning out to be exceedingly defective, he had recourse to the celebrated expedient of comparing the state of the sun in past years with the recorded prices of corn. Fully admitting the inadequacy of the criterion, he still thought that the sun being " the ultimate fountain of fertility, the subject may deserve a short investigation, especially as no other method is left for our choice." He obtained, as the upshot, partial confirmation of the surmise that " some temporary defect of vegetation" ensued upon the subsidence of solar agitation. In plainer language, food-stuffs tended to become dear when sun-spots were few and small. No signs of cyclical change could, however, be made out. The discovery of the "sun-spot period" was left to Schwabe. This admirable preliminary effort to elicit the earth's response to solar vicissitudes was denounced by Brougham as a "grand absurdity;" and the readers of the second number of the *Edinburgh Review* were assured that "since the publication of 'Gulliver's Travels,' nothing so ridiculous had ever been offered to the world!"

Herschel did not neglect the planets. His observations of Venus extended from 1777 to 1793. Their principal object was to ascertain the circumstances of the planet's rotation; but they eluded him; which, considering that they are still quite uncertain, is not surprising. He would probably have communicated nothing on the subject had he not been piqued into premature publication by Schröter's statement that the mountains of Venus rose to " four, five, or even six times the perpendicular elevation of Chimborazo." Herschel did not believe in them, and expressed his incredulity in somewhat sarcastic terms. " As to the mountains in Venus," he wrote, " I may venture to say that no eye which is not considerably better than mine, or assisted by much better instruments, will ever get a sight of them." He rightly inferred, however, the presence of an extensive atmosphere from the bending of the sun's rays so as to form much more than a semicircular rim of light to the dark disc of the planet when near inferior conjunction—that is, when approximately in a right line between us and the sun. He fully ascertained, too, the unreality of the Cytherean phantom-satellite. The irritability visible in this paper made a solitary exception to Herschel's customary geniality. It might have led to a heated controversy but for the excellent temper of Schröter's reply.

Although we may not be prepared to gainsay Herschel's dictum that " the analogy between Mars and the earth is perhaps by far the greatest in the whole solar system," we can hardly hold it to be so probable as he did that " its inhabitants enjoy a situation in many respects similar to ours." Yet the modern epoch in the physical study of Mars began with his announcement in 1784 that its white polar caps spread and shrank as winter and summer alternated in their respective hemispheres. His conclusion of their being produced by snowy depositions from " a considerable, though moderate, atmosphere," is not likely to be overthrown. He established, besides, the general permanence of the dark markings, notwithstanding minor alterations due, he supposed, to the

G

variable distribution of clouds and vapours on the planet's surface.

This vigilant "watcher of the skies" laid before the Royal Society, May 6th, 1802, his " Observations of the two lately discovered Bodies." These were Ceres and Pallas, which, with Juno and Vesta, picked up shortly afterwards, constituted the vanguard of the planetoid army. Herschel foresaw its arrival. He adopted unhesitatingly Olbers's theory of their disruptive origin, and calculated that Mercury, the least of the true planets, might be broken up into 35,000 masses no larger than Pallas. An indefinite number of such fragments (about 420 are now known) were accordingly inferred to circulate between the orbits of Mars and Jupiter. He distinguished their peculiarities, and, since they could with propriety be designated neither planets nor comets, he proposed for them the name of "asteroids." But here again he incurred, to use his own mild phrase, "the illiberal criticism of the *Edinburgh Review.*" "Dr. Herschel's passion for coining words and idioms," Brougham declared, "has often struck us as a weakness wholly unworthy of him. The invention of a name is but a poor achievement in him who has discovered whole worlds." The reviewer forgot, however, that new things will not always fit into the framework of old terminology. He added the contemptible insinuation that Herschel had devised the word "asteroid" for the express purpose of keeping Piazzi's and Olbers's discoveries on a lower level than his own of Uranus.

Herschel made no direct reply to the attack; only pointing out, in December, 1804, how aptly the detection of Juno had come to verify his forecasts. "The specific differences," he said, " between planets and asteroids appear now, by the addition of a third individual of the latter species, to be more fully established ; and that circumstance, in my opinion, has added more to the ornament of our system than the discovery of another planet could have done."

His endeavours to determine the diameters of these small bodies were ineffectual. Although he at first estimated those of Ceres and Pallas at 162 and 147 miles, he admitted later his inability to decide as to the reality of the minute discs shown by them ; and they were first genuinely measured by Professor Barnard with the great Lick refractor in 1894.

The "trade-wind theory" of Jupiter's belts originated with Herschel ; and he took note of the irregular drifting movements of the spots on his surface, and their consequent uselessness for deter- mining the period of his rotation.   That of Saturn's he fixed quite accurately at ten hours sixteen minutes, with a marginal uncertainty of two minutes, the period now accepted being of ten hours fourteen minutes.   The possession by this planet of a profound atmosphere was inferred from the changes in its belts, as well as from some curious phenomena attending the disappearance of its satellites.   They were commonly seen to " hang on the limb "—that is, to pause during an appreciable interval on the brink of occultation.   Mimas, on one occasion, remained thus poised during twenty minutes! For so long it was geometrically concealed, although visible by the effect of refraction.   Saturn was an object of constant solicitude at Slough ; and it was only with the sur- passing instruments mounted there that much could be learned about Galileo's *altissimo pianeta*. Herschel supposed, with Laplace, the rings to be solid structures ;

G 2

and he added that the interval of 2,500 miles separating them "must be of considerable service to
the planet in reducing the space that is eclipsed by
the shadow of the ring." The "crape ring" was *seen*,
but not recognised. In one of his drawings it figures
as a dusky belt crossing the body of the planet.

His satellite discoveries proved exceedingly difficult
to verify. The Saturnian pair were lost, after he left
them, until his son once more drew them from
obscurity. Regarding the outermost member of the
system, Japetus, discovered by Cassini in 1671,
Herschel noticed, in 1792, a singular circumstance. It
was already known to vary in brightness; we receive
from it, in fact, four and a-half times more light at
certain epochs than at others. The novelty consisted
in showing that this variation depended upon the
satellite's situation in its orbit in such a manner as to
leave no doubt that, like our moon, it keeps the same
face always directed inwards towards its primary. So
that Japetus was inferred to turn on its axis in the
period of its revolution round Saturn, that is, in seventy-
nine and one-third days.

"From its changes" he "concluded that by far the
largest part of its surface reflects much less light than
the rest; and that neither the darkest nor the
brightest side of the satellite is turned towards the
planet, but partly the one and partly the other."

Guessing at once that our moon and Japetus did
not present the only examples of equality in the times
of rotation and revolution, he continued: "I cannot
help reflecting with some pleasure on the discovery of
an analogy which shows that a certain uniform plan
is carried on among the secondaries of our solar
system; and we may conjecture that probably most

of the satellites are governed by the same law, especially if it be founded upon such a construction of their figure as makes them more ponderous towards their primary planet." This very explanation was long afterwards adopted by Hansen. The peculiarity in question may without hesitation beset down as an effect of primordial tides.

In 1797 Herschel brought forward detailed evidence to shew that his generalisation applied to the Jovian system ; but recent observations at Lick and Arequipa demand a suspension of judgment on the point.

The Uranian train of attendants was left by Herschel in an unsettled condition. Two of them, as we have seen, he discovered in 1787 ; and he subsequently caught glimpses of what he took to be four others. But only Oberon and Titania have maintained their status : the four companions assigned to them are non-existent. An unmistakable interior pair— Ariel and Umbriel—was, however, discovered by Mr. Lassell, at Malta, in 1851 ; and they may possibly have combined with deceptive star-points to produce Herschel's dubious quartette. He described in 1798 the exceptional arrangement of the Uranian system. Its circulation is retrograde. The bodies composing it move from east to west, but in orbits so tilted as to deviate but slightly from perpendicularity to the plane of the ecliptic.

No trifling sensation was created in 1783, and again in 1787, by the news that Herschel had seen three lunar volcanoes in violent eruption. " The appearance of the actual fire " in one of them was compared by him to " a small piece of burning charcoal when it is covered with a very thin coating of white ashes. All the adjacent parts of the volcanic mountain seemed to be faintly illuminated by the eruption, and were gradually more obscure as they lay at a greater distance from the crater." He eventually became aware that his senses had imposed upon him; but the illusion was very complete and has since occasionally been repeated. What was really seen was probably the vivid reflection of earth-shine from some unusually white lunar summits.

He never knowingly discovered a comet, although some few such bodies possibly ensconced themselves, under false pretences, in his lists of nebulæ. But he made valuable observations upon the chief of those visible in his time, and introduced the useful terms, corresponding to instructive distinctions, "head," "nucleus," and "coma." He inferred from the partial phases of the comet of 1807, that it was in a measure self-luminous; and from their total absence in the great comet of 1811, that its light was almost wholly original. The head of this object, which shone with an even, planetary radiance, he determined to be 127,000, the star-like nucleus within, 428 miles across. The tail he described as "a hollow, inverted cone," one hundred millions of miles long, and fifteen millions broad. This prodigious appurtenance was, in grade of luminosity, an exact match for the Milky Way. That comets wear out by the waste of their substance at perihelion, he thought very probable; the extent of their gleaming appendages thus serving as a criterion of their antiquity. They might, indeed, arrive in the solar system already shorn of much of their splendour by passages round other suns than ours; but their "age" could, in any case, be estimated according to the progress made in their decline from a purely nebulous to an almost " planetary " state. He went so far as to throw out the conjecture that " comets may become asteroids ; " although the converse proposition that " asteroids may become comets," of which something has been heard lately, would scarcely have been entertained by him.

Enough has been said to show how greatly knowledge of the solar system in all its parts was furthered by Herschel's observational resources, fertility of invention, and indomitable energy. He was, so to speak, ubiquitous. He had taken all the heavens for his province. Nothing that they included, from the faintest nebula to the sun, and from the sun to a telescopic shooting-star, evaded his consideration. A whole cycle of discoveries and successful investigations began and ended with him.

His fame as an astronomer has cast into the shade his merits as a physicist. He made pioneering experiments on the infra-red heat-rays,[*] and anticipated, by an admirable intuition, the fact ascertained with the aid of Professor Langley's " bolometer," that the invisible surpass in extent the visible portions of the solar spectrum.[†]   A search for darkening glasses suitable to solar observations, led him to the inquiry. Finding that some coloured media transmitted much heat and little light, while others stopped heat and let through most of the light, he surmised that a different heating power might belong to each spectral tint.   His own maxim that " it is sometimes of great use in natural philosophy to doubt of things that are commonly taken for granted," here came in appropriately.   With a free mind he set about determining the luminous and thermal powers of

* Phil. Trans. 1800, p. 255.   † *Ibid.*, p. 291.

successive spectral regions. They seemed to vary quite disconnectedly. A thermometer exposed to red rays during a given interval, rose three and a half times as much as when exposed to violet rays; and he showed further, by tracing the heat- and light-curves of the prismatic spectrum, that its heat-maximum lay out of reach of the eye in the infra-red, while luminous intensity culminated in the yellow. He even threw out the sagacious conjecture that " the chemical properties of the prismatic colours" might be "as different as those which relate to light and heat;" adding that " we cannot too minutely enter into an analysis of light, which is the most subtle of all the active principles that are concerned in the operations of nature."

The ardour with which he pursued the inquiry betrays itself in the rapid succession of four masterly essays communicated to the Royal Society in 1800. They contained the first exposition worth mentioning of the properties of radiant heat. They gave the details of experiments demonstrating its obedience to the same laws of reflection, refraction, and dispersion as light; and showing the varieties in the absorptive action upon it of different substances. In the third memoir of the series, Professor Holden finds himself at a loss " which to admire most—the marvellous skill evinced in acquiring such accurate data with such inadequate means, and in varying and testing such a number of questions as were suggested in the course of the investigation—or the intellectual power shown in marshalling and reducing to a system such intricate, and apparently self-contradictory phenomena." There is, indeed, scarcely one of Herschel's researches in which his initiative vigour and insight are more brilliantly displayed than in this *parergon*—this task executed, as it were, out of hours. It is only a pity that he felt compelled, by the incompatibility of their distribution in the spectrum, to abandon his original opinion in favour of the essential identity of light and radiant heat. The erroneous impression left on the public mind by his recantation has hardly yet been altogether effaced.

CHAPTER V.

THE INFLUENCE OF HERSCHEL'S CAREER ON MODERN
ASTRONOMY.

THE powers of the telescope were so unexpectedly
increased, that they may almost be said to have been
discovered by William Herschel. No one before him
had considered the advantages of large apertures. No
one had seemed to remember that the primary
function of an instrument designed to aid vision is to
collect light. The elementary principle of space-
penetration had not been adverted to. It devolved
upon him to point out that the distances of similar
objects are exactly proportional to the size of the
telescopes barely sufficing to show them. The reason
is obvious. Compare, for instance, a one-inch telescope with the naked eye. The telescope brings to a
focus twenty-five times as much light as can enter the
pupil, taken at one-fifth of an inch in diameter;
therefore it will render visible a star twenty-five times
fainter than the smallest seen without its help; or—
what comes to the same thing—an intrinsically equal
star at a five-fold distance. A one-inch glass hence
actually quintuples the diameter of the visible
universe, and gives access to seventy-five times the
volume of space ranged through by the unassisted eye.

This simple law Herschel made the foundation-
stone of his sidereal edifice. He was the first to
notice it, because he was the first practically to concern himself with the star-depths. The possibility of

gauging the heavens rose with him above the horizon of science. Because untiring in exploration, he was insatiable of light; and being insatiable of light, he built great telescopes.

His example was inevitably imitated and surpassed. Not through a vulgar ambition to " beat the record," but because a realm had been thrown open which astronomers could not but desire to visit and search through for themselves. Lord Rosse's six-foot reflector was the immediate successor of Herschel's four-foot; Mr. Lassell's beautiful specula followed; and the series of large *metallic* reflectors virtually closed with that of four-feet aperture erected at Melbourne in 1870. The reflecting surface in modern instruments is furnished by a thin film of silver deposited on glass. It has the advantage of returning about half as much again of the incident light as the old specula, so that equal power is obtained with less size. Dr. Common's five-foot is the grand exemplar in this kind; and it is fully equivalent to the Parsonstown six-foot.

The improvement of refractors proceeded more slowly. Difficulties in the manufacture of glass stood in the way, and difficulties in the correction of colour. The splendid success, however, of the Lick thirty-six inch, and the fine promise of the Yerkes forty-inch, have turned the strongest current of hope for the the future in the direction of this class of instrument. But all modern efforts to widen telescopic capacity primarily derive their impulse from Herschel's passionate desire to see further, and to see better, than his predecessors.

His observations demonstrate the rare excellence of his instruments. Experiments made on the asteroid

Juno, in 1805, for the purpose of establishing a valid distinction between real and fictitious star-discs, prove, in Professor Holden's opinion, the reflector employed to have been of almost ideal perfection ; and his following of Saturn's inner satellites right up to the limb, with the twenty-foot and the forty-foot, was a *tour de force* in vision scarcely, if ever, surpassed.

In the ordinary telescopes of those days really good definition was unknown ; they showed the stars with rays or tails, distorted into triangles, or bulged into " cocked hats ; " clean-cut, circular images were out of the question. Sitting next Herschel one day at dinner, Henry Cavendish, the great chemist, a remarkably taciturn man, broke silence with the abrupt question—" Is it true, Dr. Herschel, that you see the stars round ? "   " Round as a button," replied the Doctor ; and no more was said until Cavendish, near the close of the repast, repeated interrogatively, " Round as a button ? "   " Round as a button," Herschel briskly reiterated, and the conversation closed.

It seems probable that Herschel's *caput artis* lost some of its fine qualities with time.   Great specula are peculiarly liable to deterioration. Their figure tends to become impaired by the stress of their own weight; their lustre is necessarily more or less evanescent. Re-polishing, however, is a sort of re-making ; and the last felicitous touches, upon which everything depends, can never be reckoned upon with certainty. Hence, the original faultlessness of the great mirror was, perhaps, never subsequently reproduced.

" Such telescopes as Herschel worked with," Dr. Kitchiner wrote in 1815, " could only be made by the man who used them, and. only be used by the man who made them." The saying is strictly true. His skill in one branch promoted his success in the other. He was as much at home with his telescopes as the Bedouin are with their horses. Their peculiarities made part of his most intimate experience. From the graduated varieties of his specula he picked out the one best suited to the purpose in hand. It was his principle never to employ a larger instrument than was necessary, agility of movement being taken into account no less than capacity for collecting light. The time-element, indeed, always entered into his calculations; he worked like a man who has few to-morrows.

His sense of sight was exceedingly refined, and he took care to keep it so. In order to secure complete " tranquility of the retina," he used to remain twenty minutes in the dark before attempting to observe faint objects ; and his eye became so sensitive after some hours spent in "sweeping," that the approach of a third-magnitude star obliged him to withdraw it from the telescope. A black hood thrown over his head while observing served to heighten this delicacy of vision. He despised no precaution. Details are " of consequence," he wrote to Alexander Aubert, an amateur astronomer, " when we come to refinements, and want to *screw an instrument up to the utmost pitch.*"

This was said in reference to his application of what seemed extravagantly high magnifying powers. He laid great stress upon it in the earlier part of his career. The method, he said, was " an untrodden path," in which "a variety of new phenomena may be expected." With his seven-foot Newtonian he used magnifications up to nearly 6,000, proceeding, however, "all along experimentally"—a plan far too much neglected in " the art of seeing." " We are told," he proceeded, "that we gain nothing by magnifying *too much*. I grant it, but shall never believe I magnify too much till by experience I find that I can see better with a lower power." The innovation was received with a mixture of wonder, incredulity, and admiration.

Herschel showed his customary judgment in this branch of astronomical practice. He established the distinctions still maintained, and laid down the lines still followed. It is true he went far beyond the point where modern observers find it advisable to stop. The highest power brought into use with the Lick refractor is 2,600; and Herschel's instruments bore 5,800 (nominally 6,500) without injury to definition. But only at exceptional moments. His habitual sweeping power was 460; he "screwed-up" higher only for particular purposes, and under favourable conditions. Although his strong eye-pieces seem, for intelligible reasons, to have been laid aside on the adoption of the "front-view" form of construction, they had served him well in the division of close pairs, as well as for bringing faint stars into view—an effect correctly explained by him as due to the augmented darkness, under high powers, of the sky-ground. But the most important result of their employment was the discovery that the stars have no sensible dimensions. This became evident through the failure of attempts to magnify them; the higher the power applied, the smaller and more intense they appeared. Herschel accordingly pronounced stellar telescopic discs " spurious," but made no attempt to explain their origin through diffraction.

He never possessed an instrument mounted equatoreally—that is, so as automatically to follow the stars. In its absence, his work, had it not been accomplished, would have seemed to modern ideas impossible. No clockwork movement kept the objects he was observing in the field of view. His hands were continually engaged in supplying the deficiency. How, under these circumstances, he contrived to measure hundreds of double stars, and secure the places of thousands of nebulæ, would be incomprehensible but for the quasi-omnipotence of enthusiasm.

The angle made with the meridian by the line joining two stars (their "position-angle") was never thought of as a quantity useful to be ascertained until Herschel, about 1779, invented his "revolving-wire micrometer." This differed in no important respect from the modern "filar micrometer;" only spider-lines have been substituted for the original silk fibres. For measuring the distances of the wider classes of double stars, he devised in 1782 a "lamp-micrometer;" while those of the closest pairs were estimated in terms of the discs of the components. In compiling his second catalogue, however, he used the thread-micrometer for both purposes. It is true that "even in his matchless hands"—in Dr. Gill's phrase—the results obtained were "crude;" but the fact remains that the whole system of micrometrical measurement came into existence through Herschel's double-star determinations.

Their consequences have developed enormously within the last few years. Mr. Burnham's discoveries of excessively close pairs have been so numerous as to leave no reasonable doubt that their indefinite multiplication is only a question of telescopic possibility.

Then in 1889, another power came into play; the spectroscope took up the work of resolving stars. Or rather, the spectroscope in alliance with the photographic camera; for the spectral changes indicating the direction and velocity of motion in the line of sight can be systematically studied, as a rule, only when registered on sensitive plates. The upshot has been to bring within the cognisance of science the marvellous systems known as "spectroscopic binaries." They are of great variety. Some consist of a bright, others of a bright and dark, pair. Those that revolve in a plane nearly coinciding with our line of vision undergo mutual occultations. A further detachment seem to escape eclipse, yet vary in light for some unexplained reason, while they revolve. Others, like Spica Virginis, revolve without varying. Their orbital periods are counted by hours or days. The study of the disturbances of these remarkable combinations promises to open a new era in astronomical theory. For they are most likely all multiple. Irregularities indicating the presence of attractive, although obscure bodies, have, in several cases, been already noticed.

The revolutions of spectroscopic binary stars can be studied to the greatest advantage when they involve light-change; and photometric methods have accordingly begun to play an important part in the sidereal department of gravitational science. And here again we meet with Herschel's initiative. His method of sequences has been already explained; and he made the first attempt to lay down a definite scale of star-magnitudes. He failed, and it was hardly desirable that he should succeed. On his scale, the ratio of change from one grade to the next constantly diminished. In the modern system it remains always the same. A star of the second magnitude is by definition two and a-half (2·512) times less bright than one of the first; a star of the third magnitude is two and a-half times less bright than one of the second, the series descending without modification until beyond telescopic reach. This uniformity in the *proportionate* value of a magnitude is indispensable for securing a practicable standard of measurement. Herschel, however, took the great step of introducing a principle of order.

His estimates of stellar lustre were purely visual. And although various instruments, devised for the purpose, have since proved eminently useful, the ultimate appeal in all is to the eye. But there are many signs that, in the photometry of the future, not the eye but the camera will be consulted. Their appraisements differ markedly. Herschel's incidental remark on the disturbance of light-valuation by colour touches a point of fundamental importance in photographic photometry. The chemical method gives to white stars a great advantage over yellow and red ones. They come out proportionately much brighter on the sensitive plate than they appear to the eye. And to these varieties of hue correspond spectral class-distinctions, the spectrum of an object being nothing but its colour written at full length. This systematic discrepancy between visual and photographic impressions of brightness, while introducing unwelcome complications in measures of magnitude, may serve to bring out important truths. The inference, for example, has been founded upon it that the Milky Way is composed almost exclusively of white, or "Sirian" stars; and there can be no

H

question but that the arrangement of stars in space has some respect to their spectral types.

Herschel's plan of inquiry into the laws of stellar distribution by "photometric enumeration," or gauging by magnitudes, was a bequest to posterity which has been turned to account with very little acknowledgment of its source. Argelander's review of the northern heavens (lately completed photographically by Dr. Gill to the southern pole) afforded, from 1862, materials for its application on a large scale; but the magnitudes assigned to his 324,000 stars do not possess the regularity needed to make deductions based on them perfectly trustworthy. Otherwise the distance from the earth of the actual aggregations in the Milky Way could have been ascertained in a rough way from the numerical representation of the various photometric classes. As it is, the presumption is strong that the galactic clouds are wholly independent of stars brighter than the ninth magnitude—that they only begin to gather at a depth in space whence light takes *at least* a thousand years to travel to our eyes. Confirmatory evidence, published in 1894, has been supplied by M. Easton's research, based on the same principle, into the detailed relations of stars of various magnitudes to Milky Way structure. They are exhibited only by those of the ninth magnitude, or fainter ; for with them sets in a significant crowding upon its condensed parts, attended by a scarcity over its comparative vacuities. Counts by magnitudes have, besides, made it clear that the stars, in portions of the sky removed from the Milky Way, thin out notably before the eleventh magnitude is reached ; so that, outside the galactic zone, the stellar system is easily fathomed.

Also on the strength of photometric enumerations, Dr. Gould, of Boston, came to the conclusion, in 1879, that there is an extra thronging of stars about our sun, which forms one of a special group consisting of some four or five hundred members. The publication, in 1890, of the "Draper Catalogue," of 10,530 photographed stellar spectra, has thrown fresh light on this interesting subject. Mr. Monck, of Dublin, gave reasons for holding stars physically like the sun to be generally nearer to us than stars of the Sirian class; and Professor Kapteyn, of Gröningen, as the result of a singularly able investigation, concluded with much probability that the sun belongs to a strongly condensed group of mostly "solar" stars, nearly concentric with the galaxy. It might, in fact, be said that we live in a globular cluster, since our native star-collection should appear from a very great distance under that distinctive form.

This modern quasi-discovery was anticipated by Herschel. He was avowedly indebted, it is true, to Michell's "admirable idea" of the stars being divided into separate groups; but Michell did not trouble himself about the means of its possible verification, and Herschel did. He always looked round to see if there were not some touchstone of fact within reach.

His discussion of the solar cluster, though brief and incidental, is not without present interest. He found the federative arrangement of the stars to be "every day more confirmed by observation." The "flying synods of worlds" formed by them must gravitate one towards another as if concentrated at their several centres of gravity. Accordingly, "a star, or sun, such as ours, may have a proper motion within its own system of stars, while the whole may have another proper motion totally different in quantity and direction." We may thus, he continued, "arrive in process of time, at a knowledge of all the real, complicated motions of the planet we inhabit; of the solar system to which it belongs; and even of the sidereal system of which the sun may possibly be a member." He proceeded to explain how stars, making part of the solar cluster, might be discriminated from those exterior to it; the former showing the perspective influence only of the sun's translation among themselves, while the latter would be affected besides by a "still remoter parallax"—a secular drift, compounded of the proper motion of the sun within its cluster, and of its cluster relatively to other clusters.

The possibility of applying Herschel's test is now fully recognised. Each fresh determination of the solar apex is scrutinised for symptoms of the higher "systematical parallax;" although as yet with dubious or negative results. Associated stellar groups are, nevertheless, met with in various parts of the sky. Herschel not only anticipated their existence, but suggested "a concurrence of proper motions" as the fittest means for identifying them.

His anticipation has been realised by Mr. Proctor's detection of "star-drift." Several stars in the Plough thus form a squadron sailing the same course; and similar combinations, on an apparently smaller scale, have been pieced together in various constellations. But the principle of their connection has yet to be discovered. They are evidently not self-centred systems; hence their companionship, however prolonged, must finally terminate. The only pronounced cluster with a common proper motion is the Pleiades;

and its drift seems to be merely of a perspective nature—a reflection of the sun's advance.

Bessel said of Herschel that "he aimed at acquiring knowledge, not of the motions, but of the constitution of the heavenly bodies, and of the structure of the sidereal edifice." This, however, is a defective appreciation. He made, indeed, no meridian observations, and computed no planetary or cometary perturbations; yet if there ever was an astronomer who instinctively "looked before and after," it was he. Could he have attained to a complete knowledge of the architecture of the heavens, as they stood at a given moment, it would not have satisfied him. To interpret the past and future by the present was his constant aim; from his "retired situation" on the earth, he watched with awe the grand procession of the sum of things defile through endless ages. He could not observe what was without at the same time seeking to divine what had been, and to forecast what was to come.

His nebular theory is now accepted almost as a matter of course. The spectroscope has lent it powerful support by proving the *de facto* existence of the "lucid medium," postulated by him as a logical necessity. This was done August 1st, 1864, when Dr. Huggins derived from a planetary nebula in Draco a spectrum characteristic of a gaseous body, because consisting of bright lines. Their wave-lengths, which turned out to be identical for all objects of the kind, with one or two possible exceptions, indicated a composition out of hydrogen mixed with certain unfamiliar aeriform substances. Herschel's visual discrimination of gaseous nebulæ was highly felicitous. Modern science agrees with him in

pronouncing the Orion nebula, as well as others of
the irregular class, planetaries, diffused nebulosities,
and the "atmospheres" of "cloudy stars," to be
masses of "shining fluid." As for his "ambiguous
objects," they remain ambiguous still. "Clusters in
disguise" through enormous distance, give apparently
the same quality of light with irresolvable nebulæ.
His inference that stars and nebulæ form mixed
systems has, moreover, been amply confirmed. No
one now denies their significant affinity, and very few
their genetic relationship.

Herschel gave a list in 1811 of fifty-two dim,
indefinite nebulosities, covering in the aggregate 152
square degrees. "But this," he added, "gives us by
no means the real limits" of the luminous appearance ;
"while the depth corresponding to its superficial extent
"may be far beyond the reach of our telescopes ;"
so "that the abundance of nebulous matter diffused
through such an expansion of the heavens must
exceed all imagination."

"The prophetic spirit of these remarks," Professor Barnard comments, "is being every day
made more evident through the revelations of photo-
graphy." He is himself one of the very few who have
telescopically verified any part of these suggestive
observations.

"I am familiar," he wrote in *Knowledge*, January,
1892, "with a number of regions in the heavens
where vast diffusions of nebulous matter are situated.
One of these, in a singularly blank region, lies some
five or six degrees north-west of Antares, and covers
many square degrees. Another lies north of the
Pleiades, between the cluster and the Milky Way ;
a portion of this has recently been successfully photo-

graphed by Dr. Archenhold. There is a large nebulous spot in that region, easily visible to the naked eye, which I have seen for many years. When sweeping there with a low power, the whole region between the Pleiades and the Milky Way is perceived to be nebulous. These great areas of nebulosity make their presence known by a singular dulling of the ordinarily black sky, as if a thin veil of dust intervened." They "are specially suitable for the photographic plate, and it is only by such means that they can be at all satisfactorily located."

Some of Herschel's milky tracts have been thus pictured; notably one in the Swan, shown on Dr. Max Wolf's plates to involve the bright star Gamma Cygni; and another immense formation extending over sixty square degrees about the belt and sword of Orion, and joining on, Herschel was "pretty sure," to the great nebula. This, never unmistakably *seen* except by him, portrayed itself emphatically in 1886 in Professor E. C. Pickering's photographs. Herschel's persuasion of the subordinate character of the original "Fish-mouth nebula" was well-grounded. On plates exposed by Professors W. H. Pickering and Barnard, it is disclosed as the mere nucleus of a tremendous spiral, sweeping round from Bellatrix to Rigel.

Diffused nebulosities appear in photographs as far from homogeneous. They are not simple volumes of gas indefinitely expanding in all directions, after the manner of simple aeriform fluids. They possess, on the contrary, characteristic shapes. Structureless nebulæ, like structureless protoplasm, seem to be non-existent. In all, an organising principle is at work.

Minute telescopic stars showed to Herschel as

prevalently red, owing, he conjectured, to the enfeeble-
ment of their blue rays during an uncommonly
long journey through space "not quite destitute of
some very subtle medium." The argument is a
remarkable one. It would be valid if the ethereal
vehicle of light exercised absorption after the manner
of ordinary attenuated substances. There is, however,
reason to suppose that the symptomatic redness was
only a subjective impression, not an objective fact.
His colour-sense was not quite normal. The lower,
to his perception, somewhat overbalanced the higher
end of the spectrum, and his mirrors added to the
inequality by reflecting a diminished proportion of
blue light. Thus he recorded many stars as tinged
with red which are now colourless, yet lie under no
suspicion of change.

Herschel was, in the highest and widest sense, the
founder of sidereal astronomy. He organised the
science and set it going; he laid down the principles
of its future action; he accumulated materials for its
generalisations, and gave examples of how best to
employ them. His work was at once so stimulating
and so practical that its abandonment might be called
impossible. Others were sure to resume where he
had left off. His son was his first and fittest
successor; he was the only one who undertook in its
entirety the inherited task. Yet there are to be found
in every quarter of the world men imbued with
William Herschel's sublime ambitions. Success swells
the ranks of an invading army; and the march of
astronomy has, within the last decade, assumed a
triumphal character. The victory can never be completely won; the march can never reach its final goal;
but spoils are meanwhile gathered up by the wayside

which eager recruits are crowding in to share. The heavens are, year by year, giving up secrets long and patiently watched for, while holding in reserve many others still more mysterious. There is no fear of interest being exhausted by disclosure.

Herschel's dim intuition that something might be learned about the physical nature of the stars from the diverse quality of their light, was verified after sixty-five years, by the early researches of Secchi, Huggins, and Miller; but he could not suspect that, through the chemical properties, which he guessed to belong in varying degrees to the different sections of their spectra, pictures of the heavenly bodies would be obtained more perfect than the telescopic views he rapturously gazed at. Still less could he have imagined that, owing to its faculty of accumulating impressions too weak to affect the eye separately, the chemical would, in great measure, supersede the telescopic method in carrying out the designs he had most at heart.

Those designs have now grown to be of international importance. At eighteen northern and southern observatories a photographic review of the heavens is in progress. The combined results will be the registration, in place and magnitude, of fifteen to twenty millions of stars. The gauging of the skies will then be complete down to the fourteenth magnitude; and the "construction of the heavens" can be studied with materials of the best quality, and almost indefinite in quantity. By simply "counting the gauges" on Herschel's early plan, much may be learnt; the amount of stellar condensation towards the plane of the Milky Way, for instance, and the extent of stellar denudation near its poles. A marked contrast between the measures of distribution in these opposite directions will most likely be brought into view. The application of his later method of enumeration by magnitudes ought to prove even more instructive, but may be very difficult. The obstacles, it is to be hoped, will not be insurmountable; yet they look just now formidable enough.

The grand problem with which Herschel grappled all his life involves more complicated relations than he was aware of. It might be compared to a fortress, the citadel of which can only be approached after innumerable outworks have been stormed. That one man, urged on by the exalted curiosity inspired by the contemplation of the heavens, attempted to carry it by a *coup de main*, and, having made no inconsiderable breach in its fortifications, withdrew from the assault, his "banner torn, but flying," must always be remembered with amazement.

CAROLINE LUCRETIA HERSCHEL.

(From a portrait taken by Tielemann in 1829.)

CHAPTER VI.

CAROLINE HERSCHEL.

CAROLINE LUCRETIA HERSCHEL was born at Hanover, March 16th, 1750, and was thus more than eleven years younger than the brother with whose name hers is inseparably associated. She remembered the panic caused by the earthquake of 1755, and her experience barely fell short of the political earthquake of 1848; but the fundamental impressions of her long life were connected with "minding the heavens."

She was of little account in her family, except as a menial. Her father, indeed, a man of high character and cultivated mind, thought much of her future, and wished to improve her prospects by giving her some accomplishments. So he taught her to play the violin well enough to take part in concerted music. But her instruction was practicable only when her mother was out of the way, or in a particularly good humour. Essentially a "Hausfrau," Anna Ilse had no sympathy with aspirations. She was hard-working and well-meaning, but narrow and inflexible, and she kept her second daughter strictly to household drudgery. Her literary education, accordingly, got no farther than reading and writing; even the third " R " was denied to her. But she was carefully trained in plain sewing and knitting, and supplied her four brothers with stockings from so early an age that the first specimen of her workmanship touched the ground while she stood upright finishing the toe ! Few signs of tenderness were accorded to her. Her eldest brother, Jacob, a brilliant musician, and somewhat high-and-mighty in his ways, did not spare cuffs when she waited awkwardly at table; and her sister, Mrs. Griesbach, evidently took slight notice of her. William, however, showed her invariable affection; and him and her father she silently adored. In 1756, when they both returned from England with the Hanoverian Guard, she recalled how, on the day of their arrival,

" My mother being very busy preparing dinner, had suffered me to go all alone to the parade to meet my father, but I could not find him anywhere, nor anybody whom I knew; so at last, when nearly frozen to death, I came home and found them all at table. My dear brother William threw down his knife and fork, and ran to welcome, and crouched down to me, which made me forget all my grievances. The rest were so happy at seeing one another again that my absence had never been perceived."

How well one can realise the disconsolate little expedition, the woe-begone entry of the six-year-old maiden, her heart-chill on finding herself forgotten, and the revulsion of joy at her soldier-brother's cordial greeting !

Isaac Herschel died March 22nd, 1767. He had never recovered the campaign of Dettingen, yet struggled, in spite of growing infirmities, to earn a livelihood by giving lessons and copying music. His daughter was thrown by his loss into a "state of stupefaction," from which she roused herself, after some weeks, to consider the gloomy outlook of her destiny. She was seventeen, and was qualified, as she reflected with anguish, only to be a housemaid. She was plain in face and small in stature, and her father had often warned her that if she ever married it would be comparatively late in life, when her fine character had unfolded its attractions. Still, she did not lose hope of making her way single-handed. Although over-burthened with servile labours, she contrived, unknown to her mother, to get some teaching in fancy-work from a consumptive girl whose cough from across the street gave the signal for a daybreak rendezvous; trusting that, with this acquirement, and "a little notion of music, she might obtain a place as governess in some family where the want of a knowledge of French would be no objection." There was "no kind of ornamental needlework, knotting, plaiting hair, stringing beads and bugles, of which she did not make samples by way of mastering the art." She was then permitted to take some lessons in dressmaking and millinery. But the current of her thoughts was completely changed by an invitation from her brother William to join him at Bath. She was, if possible, to be made into a concert-singer. Yet her voice had never been tried, and its very existence was problematical. It may, then, be suspected that Willliam's primary motive was to come to the rescue of his poor little Cinderella sister.

Months passed in "harassing uncertainty" as to whether she was to go or stay; months, too, during which her own mind was divided between the longing to follow her rising star, and a certain compunctious clinging to her duties at home. Time, however, did not pass in idleness. Taking no notice of the superior Jacob's ridicule of her visionary transformation into an artist, she quietly set about practising, with a gag between her teeth, the solo parts of violin concertos, "shake and all," so that, as she says, "I had gained a tolerable execution before I knew how to sing." She occupied herself besides in making a store of prospective clothing for relatives, who, she could not but fear, would miss her services. For her withdrawal her mother, however, received from William money-compensation, which enabled her to keep a servant in lieu of her daughter. The parting, when he came to fetch her, in August, 1772, was none the less a sorrowful one; but Caroline had much to distract her mind from dwelling on those she had left behind. She had, besides, much discomfort to endure. Six days and nights in an open stage-carriage were followed by a tempestuous passage; the packet in which they embarked at Helvoetsluys reached Yarmouth dismasted and half-wrecked; and they were finally, not duly landed, but "thrown like balls by two sailors," on the English coast. After a brief glimpse of London, they started, August 28th, in the night coach for Bath, where Caroline arrived "almost annihilated" by fatigue and want of sleep.

Her training for an unfamiliar life began without delay. She had to learn English, arithmetic, and enough of account-keeping to qualify her for conducting the household affairs; a routine of singing-lessons and practising was entered upon; and she was sent out alone to market, Alexander Herschel lurking behind to see that she came safely out of the *mêlée* of buyers and sellers, whence she brought home "whatever in her fright she could pick up." She suffered many things, too, from her brother's servant, "a hot-headed old Welshwoman," whose *régime* was one of rack and ruin to domestic utensils; while *heimweh* made formidable onslaughts on her naturally serene spirits.

A visit to London, as the guest of Mrs. Colebrook,

one of her brother's pupils, gave her some experience of town gaieties. But the expenses of dress and chairmen shocked her frugal ideas; and she thought the young ladies, whose companionship was offered to her, "very little better than idiots." As a vocalist, Miss Herschel came easily to the front. After a few months of study, her voice was in demand at evening parties; when her foreign accentuation had been corrected, she took the first soprano parts in "The Messiah," "Samson," "Judas Maccabæus," and other oratorios; and sang as prima donna at the winter concerts both at Bath and Bristol. In accordance with her resolution to appear only where her brother conducted, she declined an engagement for a musical festival at Birmingham. A year's training in deportment was a preliminary to her *début*; a celebrated dancing mistress being engaged—to use Caroline's own phrase—" in drilling me for a gentlewoman. Heaven knows how she succeeded!" A gift of ten guineas from William provided her with a dress which made her, she was told, "an ornament to the stage;" and she was complimented by the Marchioness of Lothian on " pronouncing her words like an Englishwoman." Her success was decided, and promised to be enduring enough to satisfy her modest ambition of supporting herself independently.

It was, however, balked by an extraordinary turn of affairs; a turn at first not at all to her liking. After the lapse of half-a-century she still set it down as the grievance of her life that " I have been throughout annoyed and hindered in my endeavours at perfecting myself in any branch of knowledge by which I could hope to gain a creditable livelihood."

William Herschel, when Caroline joined him at

Bath, was just feeling his way towards telescope-making. The fancy did not please her. The beginnings of great things are usually a disturbance and an anxiety. They imply a draft upon the future which may never be honoured, and they often play sad havoc with the present. And Miss Herschel was business-like and matter-of-fact. But her devotion triumphed over her common-sense. Keeping her misgivings to herself, she met unlooked-for demands with the utmost zeal, intelligence, and discretion. She was always at hand when wanted, yet never in the way. Through her care, some degree of domestic comfort was maintained amid the unwonted confusion of optical manufacture. During the tedious process of mirror-polishing, she sustained her brother physically and mentally, putting food into his mouth, and reading aloud "Don Quixote," and the "Arabian Nights." She was ready with direct aid, too, and "became in time as useful a member of the workshop as a boy might be to his master in the first year of his apprenticeship." "Alex," she continued, "was always very alert, assisting when anything new was going forward ; but he wanted perseverance, and never liked to confine himself at home for many hours together. And so it happened that my brother William was obliged to make trial of my abilities in copying for him catalogues, tables, and sometimes whole papers which were lent him for his perusal." Musical business, meantime, received due attention. Steady preparation was made for concerts and oratorios ; choruses were instructed, rehearsals attended, parts diligently written out from scores. But the discovery of Uranus swept away the necessity for these occupations ; and with a final performance in St.

Margaret's Chapel, on Whit-Sunday, 1782, the musical career of William and Caroline Herschel came to a close.

Miss Herschel's " thoughts were anything but cheerful " on the occasion. She saw the terrestrial ground cut from under her feet, and did not yet appreciate the celestial situation held in reserve for her. Music, in her opinion, was her true and only vocation ; the contemplation of herself in the guise of an assistant-astronomer moved her to cynical self-scorn. As usual, however, her personal wishes were suppressed. Housewifely cares, too, weighed upon her. The dilapidated gazebo at Datchet provided no suitable shelter for a well-regulated establishment. It was roofed more in appearance than in reality ; the plaster fell from the ceilings ; the walls dripped with damp ; rheumatism and ague were its rightful inmates. Then the prices of provisions appalled her, especially in view of the scarcity of five-pound notes since the opulence of Bath had been exchanged for the penury of a court precinct.

Yet she set to work with a will to learn all that was needful for her untried office. Not out of books. " My dear brother William," she wrote in '1831, " was my only teacher, and we began generally where we should have ended ; he supposing I knew all that went before." The lessons were of the most desultory kind. They consisted of answers to questions put by her as occasions arose, during breakfast, or at odd moments. The scraps of information thus snatched were carefully recorded in her commonplace book, where they constituted a miscellaneous jumble of elementary formulæ, solutions of problems in trigonometry, rules for the use of tables of logarithms, for converting sidereal into solar time, and the like.

I

Nothing was entrusted to a memory compared by her instructor to "sand, in which everything could be inscribed with ease, but as easily effaced." So that even the multiplication table was carried about in her pocket. She appears never to have spent a single hour in the systematic study of astronomy. Her method was that in vogue at Dotheboys Hall, to "go and know it," by practising, as it were, blindfold, what she had been taught. Yet a computational error has never, we believe, been imputed to her; and the volume of her work was very great.

Its progress was diversified by more exciting pursuits. She began, in 1782, to "sweep for comets," and discovered with a 27-inch reflector, in the autumn of 1783, two nebulæ of first-rate importance—one a companion to the grand object in Andromeda, the other a superb elliptical formation in Cetus. She was by this time more than reconciled to her astronomical lot; Von Magellan, indeed, reported in 1785, that brother and sister were equally captivated with the stars.

The original explorations, in which she was beginning to delight, were interrupted by the commencement of his with the "large twenty-foot." Her aid was indispensable, and from December, 1783, she "became entirely attached to the writing-desk." She was no mere mechanical assistant. A wound-up automaton would have ill served William Herschel's turn. He wanted "a being to execute his commands with the quickness of lightning"; and his commands were various. For he was making, not following precedents, and fresh exigencies continually arose. Under these novel circumstances, his sister displayed incredible zeal, promptitude, and versatility. She

would throw down her pen to run to the clock, to
fetch and carry instruments, to measure the ground
between the lamp-micrometer and the observer's eye;
discharging these, and many .other successive tasks
with a rapidity that kept pace with his swift proceedings. Fatigue, want of sleep, cold, were disregarded;
and although nature often exacted next day penalties
of weariness and depression for those nights of intense
activity, the faithful amanuensis never complained.
" I had the comfort," she remarked simply, " to see that
my brother was satisfied with my endeavours to assist
him." The service was not unaccompanied by danger.
One night poor Caroline, running in the dark over
ground a foot deep in melting snow, in order to make
some alteration in the movement of the telescope, fell
over a great hook, which entered her leg so deeply
that a couple of ounces of her flesh remained behind
when she was lifted off it. The wound was formidable
enough, in Dr. Lind's opinion, to entitle a soldier to
six weeks' hospital-nursing, but it was treated cursorily
at Datchet; the patient consoling herself for a few
nights' disablement with the reflection that her
brother, owing to cloudy weather, " was no loser
through the accident."

Busy days succeeded watchful nights. From the
materials collected at the telescope, she formed
properly arranged catalogues, calculating, in all, the
places of 2,500 nebulæ. She brought the whole of
Flamsteed's *British Catalogue*—then the *vade me-
cum* of astronomers—into zones of one degree wide,
for the purpose of William's methodical examination;
copied out his papers for the Royal Society; kept the
observing-books straight, and documents in order.
Then, in the long summer months, when " there was

I 2

nothing but grinding and polishing to be seen," she took her share of that too, and " was indulged with the last finishing of a very beautiful mirror for Sir William Watson."

On August 1st, 1786, her brother's absence leaving her free to observe on her own account, she discerned a round, hazy object, suspiciously resembling a comet. Its motion within the next twenty-four hours certified it as such, and she immediately announced the apparition to her learned friends, Dr. Blagden and Mr. Aubert. The latter declared in reply, " You have immortalised your name," and saw in imagination " your wonderfully clever and wonderfully amiable brother shedding," upon receipt of the intelligence, "a tear of joy." This was the first of a series of eight similar discoveries, in five of which her priority was unquestioned. They were comprised within eleven years, and were made, after 1790, with an excellent five-foot reflector mounted on the roof of the house at Slough. Considering that she swept the heavens only as an interlude to her regular duties, never for an hour forsaking her place beside the great telescopes in the garden, her aptitude for that fascinating pursuit must be rated very high. It was not until 1819 that Encke identified her seventh comet—detected November 7th, 1795—with one previously seen by Méchain in January, 1786. None other revolves so quickly, its returns to perihelion occurring at intervals of three and a quarter years. It has earned notoriety, besides, by a still unexplained acceleration of movement.

Caroline Herschel was the first woman to discover a comet; and her remarkable success in what Miss Burney called " her eccentric vocation," procured for her an European reputation. But the homage which she received did not disturb her sense of subordination. " Giving the sum of more to that which hath too much," she instinctively transferred her meed of praise to her brother. She held her comets, notwithstanding, very dear. All the documents relating to them were found after her death neatly assorted in a packet labelled " Bills and Receipts of my Comets "; and the telescopes with which they had been observed ranked among the chief treasures of her old age. She presented the smaller one before her death to her friend Mr. Hausmann; the five-foot to the Royal Astronomical Society, where it is religiously preserved.

The " celebrated comet-searcher" was described by Miss Burney in 1787 as "very little, very gentle, very modest, and very ingenuous; and her manners are those of a person unhackneyed and unawed by the world, yet desirous to meet and return its smiles." To Dr. Burney, ten years later, she appeared " all shyness and virgin modesty "; while Mrs. Papendick mentions her as " by no means prepossessing, but an excellent, kind-hearted creature." She was, in 1787, officially appointed her brother's assistant, with a salary of fifty pounds a year; " and in October," she relates, " I received twelve pounds ten, being the first quarterly payment of my salary, and the first money I ever in all my lifetime thought myself to be at liberty to spend to my own liking." The arrangement was made in anticipation of her brother's marriage, when —to quote her one bitter phrase on the subject— " she had to give up the place of his housekeeper." She did not readily accommodate herself to the change ; and a significant gap of ten years in her journal suggests that she wrote much during that time of struggle which her calmer judgment counselled her to destroy. Her strong sense of right and habitual abnegation, however, came to her aid; the family relations remained harmonious; and she eventually became deeply attached to her gentle sister-in-law. But from 1788 onwards, she lived in lodgings, either at Slough or Upton, whence she came regularly to the observatory to do her daily or nightly work.

Miss Herschel began in 1796, and finished in about twenty months, an Index to Flamsteed's observations of the stars in the "British Catalogue." A list of "errata" was added, together with a catalogue of 561 omitted stars. The work, one of eminent utility, was published in 1798, at the expense of the Royal Society. In August, 1799, she paid a visit to the Astronomer Royal, with the object of transcribing into his copy of Flamsteed's Observations some memoranda upon them made by her brother. "But the succession of amusements," we hear, "left me no alternative between contenting myself with one or two hours' sleep per night during the six days I was at Greenwich, and going home without having fulfilled my purpose." Needless to say that she chose the former.

The Royal family paid her many attentions, partly, no doubt, because of her intimacy with one of the ladies-in-waiting to the Queen. This was Madame Beckedorff, who although of "gentle" condition, had attended the same dressmaking class with the band-master's daughter at Hanover, in 1768. The distant acquaintanceship thus formed developed, at Windsor, into a firm friendship, transmitted in its full cordiality to a second generation An entry in Caroline's Diary tells of a dinner at Madame Beckedorff's, February

23, 1801, when the "whole party left the dining-room on the Princesses Augusta and Amelia, and the Duke of Cambridge coming in to see me." In May, 1813, during a visit to London, she passed several evenings at Buckingham House, "where I just arrived," she says, on May 12, "as the Queen and the Princesses Elizabeth and Mary, and the Princess Sophia Matilda of Gloucester, were ready to step into their chairs, going to Carlton House, full dressed for a fête, and meeting me in the hall, they stopped for near ten minutes, making each in their turn the kindest inquiries how I liked London, etc. On entering Mrs. Beckedorff's room, I found Madame D'Arblay (Miss Burney), and we spent a very pleasant evening."

Such Royal condescensions were frequent, and on occasions inconvenient. The Princesses Sophia and Amelia, in especial, took a strong liking for Miss Herschel's conversation, and often required her attendance for many hours together. She was graciously singled out for notice at the Frogmore assemblages, and became quite inured to the reception at Slough of dignitaries and *savants*. Nothing deranged the simple composure of her deportment. One would give much to know what were her private impressions about the notabilities who crossed her path; but her memoranda are, in this respect, perfectly colourless. Names and dates are jotted down with the same brevity as her entries of "work done." Even the personal troubles of years are curtly disposed of. Her brother Dietrich's stay in England from 1809 to 1813, left her not a day's respite from accumulated trouble and anxiety. Yet it occasioned only one little outburst, penned long afterwards.

" He came," she wrote, " ruined in health, spirit, and fortune, and, according to the old Hanoverian custom, I was the only one from whom all domestic comforts were expected. I hope I acquitted myself to everybody's satisfaction, for I never neglected my eldest brother's business " (Jacob Herschel died in 1792), " and the time I bestowed on Dietrich was taken entirely from my sleep, or what is generally allowed for meals, which were mostly taken running, or sometimes forgotten entirely. But why think of it now ? "

Her later journal is overshadowed with the fear of coming bereavement. Recurrences to the state of William's health become ominously frequent. " He is not only unwell, but low in spirits," she notes in February, 1817 ; and the following account of his departure for Bath, April 2, 1818, betrays her deep trouble :—

" The last moments before he stepped into the carriage were spent in walking through his library and workrooms, pointing with anxious looks to every shelf and drawer, desiring me to examine all, and to make memorandums of them as well as I could. He was hardly able to support himself, and his spirits were so low, that I found difficulty in commanding my voice so far as to give him the assurance he should find on his return that my time had not been misspent."

" May 1st.—But he returned home much worse than he went, and for several days hardly noticed my handiworks."

His last note to her, indited with an uncertain hand on a discoloured slip of paper, July 4, 1819, she put by with the inscription : " I keep this as a relic. Every line *now* traced by the hand of my dear brother becomes a treasure to me."

"Lina," it ran, "there is a great comet. I want you to assist me. Come to dine and spend the day here. If you can come soon after one o'clock we shall have time to prepare maps and telescopes. I saw its situation last night—it has a long tail."

Through that long tail the earth had, eight days previously—according to Olbers's calculations—cut its way; but the proposed observations at Slough, if made, were never published.

In October, 1821, Caroline Herschel wrote this melancholy "Finis" to what seemed to herself the only part of her life worth living. "Here closed my day-book; for one day passed like another, except that I, from my daily calls, returned to my solitary and cheerless home with increased anxiety for each following day."

Eighteen months after her loss of "the dearest and best of brothers," she at last gathered fortitude to put on paper her recollections of the "heartrending occurrences" witnessed by her during the closing months of her fifty years' sojourn in England. In every line of what she then wrote, her absorbed fidelity to him, growing more and more tenacious as the end drew visibly nigher, comes out with unconscious pathos. The anguish with which she watched each symptom of decay seared her heart, but was refused any outward expression. She played out her rôle of self-suppression until the curtain fell. A last gleam of hope visited her July 8th, 1822, when she marked down in an almanac the cheering circumstance that her invalid had "walked with a firmer step than usual above three or four times the distance from the dwelling-house to the library, in order to gather and eat raspberries in his garden with me. But," she added sadly, "I never saw the like again."

In the impetuosity of her grief, she made an irreparable mistake. Only a month earlier she had surrendered to her impecunious brother Dietrich her little funded property of £500 ; now, without reflecting on the consequences, she "gave herself, with all she was worth, to him and his family." She was in her seventy-third year; her only remaining business in life, it seemed to her, was to quit it; the virtual close of her career had come; the actual close could not long be delayed. So she retired to her native place to die promptly, if that might be, but, at any rate, to mark the chasm that separated her from the past. She soon recognised, however, that she had taken a false step. "Why did I leave happy England?" was the cry sometimes on her lips, always in her heart, for a quarter of a century. She was taken aback by her own vitality. She found out too late that her powers of work, far from being exhausted, might have been turned to account for her nephew as they had been for her brother. And it was to him and his mother, after all, that her strong affections clung. Her relatives in Hanover, although they treated her with consideration, were hopelessly uncongenial. "From the moment I set foot on German ground," she said, "I found I was alone.' Fifty years is a huge gap in a human life. Miss Herschel had been all that time progressing from the starting point where they had remained stationary. Their tastes were then necessarily incongruous with hers; nor could her interests be transplanted at will from the soil in which they were rooted. She was unable to perceive that the change was in herself. The " solitary and useless life " she led resulted, she was convinced, from her "not finding Hanover, or anyone in it, like what I left when the best of brothers took me with him to England in August, 1772!"

An exile in her old home, she felt pledged to remain there. She would not "take back her promise." For a person of her frugal habits, she was well off. Her pension of fifty pounds would have supplied her small wants, and she was reluctantly compelled to accept the annuity of £100, left to her by her brother. And since she was most generous in the bestowal of her spare cash, her presence was of some material advantage to a poor household. It gave them credit, too ; and notwithstanding that they " never could agree " in opinions, she faithfully nursed Dietrich Herschel until his death in January, 1827.

"I am still unsettled," she wrote to her nephew, December 26th, 1822, " and cannot get my books and papers in any order, for it is always noon before I am well enough to do anything, and then visitors run away with the rest of the day till the dinner-hour, which is two o'clock. Two or three evenings in each week are spoiled by company. And at the heavens there is no getting, for the high roofs of the opposite houses. But within my room I am determined nothing shall be wanting that can please my eye. Exactly facing me is a bookcase placed on a bureau, to which I will have some glass doors made, so that I can see my books. Opposite this, on a sofa, I am seated, with a sofa-table and my new writing-desk before me ; but what good I shall do there the future must tell."

Seated at that " new desk," she completed her most important work. This was the reduction into a catalogue, and the arrangement into zones, of all Sir

William Herschel's nebulæ and clusters.   Despatched to Sir John Herschel in April, 1825, it made his review of those objects feasible.   From it, he drew up his " working-list " for each night's observations ; and from it, in constructing his " General Catalogue " of 1864, he took the places of such nebulæ as he had not been able to examine personally.   In the course of the needful comparisons, " I learned," he said, "fully to appreciate the skill, diligence, and accuracy which that indefatigable lady brought to bear on a task which only the most boundless devotion could have induced her to undertake, and enabled her to accomplish." For its execution, the Gold Medal of the Royal Astronomical Society was awarded to her in 1828—an honour by which she was "more shocked than gratified." Her " Zone-Catalogue " was styled by Sir David Brewster " an extraordinary monument of the unextinguished ardour of a lady of seventy-five in the cause of abstract science."

In 1835, she was created an honorary member of the Royal Astronomical Society, Mrs. Somerville being associated with her in a distinction never before or since conferred upon a woman.   Three years later, she was surprised by the news that the Royal Irish Academy had similarly enrolled her.   " I cannot help," she wrote, " crying out aloud to myself, every now and then, What is that for ? "   The arrival, on another occasion, of presentation-copies of Mrs. Somerville's " Connexion of the Physical Sciences," and of Baily's " Account of Flamsteed," agitated her painfully.   " Coming to *me* with such things," she exclaimed, " an old, poor, sick creature in her dotage."   " I think it is almost mocking me," she added in 1840, " to look upon me as a Member of an Academy ; I that have lived these eighteen years without finding as much as a single comet."

Her local celebrity, nevertheless, diverted her. It struck her as a capital joke that she was "stared at for a learned lady." Down to 1840 she regularly attended plays and concerts, and rarely left the theatre without a "*Wie geht's?*" from His Majesty. And to find herself—"a *little* old woman"—conspicuous in the crowd, produced a sense of exhilaration. Her presence or absence was a matter of public concern, and she very seldom appeared otherwise than alert and cheerful. When close upon eighty her "nimbleness in walking," she remarked, "has hitherto gained me the admiration of all who know me; but the good folks are not aware of the arts I make use of, which consist in never leaving my room in the daytime except I am able to trip it along as if nothing were the matter." Music gave her unfailing pleasure. She heard Catalani in 1828; shared in the Paganini *furore* of 1831, and conversed with him through an interpreter. With Ole Bull she was "somewhat disappointed," finding his performance "more like conjuration than playing on a violin."

But her "painful solitude" was most of all cheered by the visits and communications of eminent men. No one of distinction in science came to Hanover without calling upon her. Humboldt, Gauss, Mädler, Encke, Schumacher, paid her their respects, personally or by letter, if not in both ways. "Next to listening to the conversation of learned men," she told the younger Lady Herschel, "I like to hear about them; but I find myself, unfortunately, among beings who like nothing but smoking, big talk on politics, wars, and such-like things." Her situation remained, to

the end, displeasing to her.   She made the best of it;
but the best was, to her thinking, bad.   Having
wilfully flung herself out of the current of life, she
was nevertheless surprised at being stranded.   She
recurred, with inextinguishable pain, to the crippling
effects of circumstances and old age.

"I lead a very idle life," she wrote in 1826.   "My
sole employment consists in keeping myself in good
humour, and not being disagreeable to others."   And
in 1839: "I get up as usual, every day, change my
clothing, eat, drink, and go to sleep again on the sofa,
except I am roused by visitors; then I talk till I can
talk no more—nineteen to the dozen!"   While at
nights "the few, few stars I can get at out of my
window only cause me vexation, for to look for the
small ones on the globe my eyes will not serve me
any longer."

She followed, however, with intense delight the
progress of her nephew's career, in which she beheld
the continuation of his father's.   The intelligence of
his having opened a nebular campaign in 1825, was
like the sound of the trumpet to a disabled war-horse.
Nothing but the decline of her powers, she assured
him, would have prevented her "coming by the first
steamboat to offer you the same assistance as, by your
father's instructions, I was enabled to afford him."
And again, in 1831: "You have made me completely
happy with the account you sent me of the double
stars; but it vexes me more and more that in this
abominable city there is no one who is capable
of partaking in the joy I feel on this revival of
your father's name.   His observations on double
stars were, from first to last, the most interesting
subject; he never lost sight of it.   And I cannot

help lamenting that he could not take to his grave the satisfaction I feel at seeing his son doing him such ample justice by endeavouring to perfect what he could only begin."

Sir John's trip to the Cape roused her ardent sympathy. "Ja!" she exclaimed, on hearing of the project, "If I were thirty or forty years younger, and could go too. In Gottes Namen!" But she was eighty-two, and could only give vent to her feelings by "jingling glasses with Betty" after dinner on his birthday, while mistress and maid together cried, "Es lebe Sir John! Hoch! Hurrah!" The reports of his achievements in the southern hemisphere were, she said, "like a drop of oil supplying my expiring lamp." "At first, on reading them," she wrote to Lady Herschel, "I could turn wild; but this is only a flash; for soon I fall into a reverie on what my dear nephew's father would have felt if such letters could have been directed to him, and cannot suppress my wish that *his* life instead of *mine* had been spared until this present moment."

The joyful intelligence of her nephew's safe return to England was sent to Miss Herschel by the Duke of Cambridge, whose attentions to her were unfailing; and she lived to hold in her hands the volume of "Cape Results," by which her brother's great survey of the heavens was rounded off to completion. But by that time the lassitude of approaching death was upon her.

Three visits from her nephew broke the monotony of separation. In October, 1824, he stopped at Hanover on his way homeward from the Continent. Before his arrival, her "arms were longing to receive him"; after his departure, she "followed him in idea every inch he moved farther " away from her. Six years passed, and then he came again.

"I found my aunt," he reported, June 19th, 1832, "wonderfully well, and very nicely and comfortably lodged, and we have since been on the full trot. She runs about the town with me, and skips up her two flights of stairs as light and fresh at least as some folks I could name who are not a fourth part of her age. In the morning till eleven or twelve she is dull and weary; but as the day advances she gains life, and is quite 'fresh and funny' at ten or eleven p.m., and sings old rhymes, nay, even dances! to the great delight of all who see her."

Their final meeting was in 1838, when Sir John's Cape laurels were just gathered; and he brought with him his eldest son, aged six. But the old lady was terrified lest the child should come to harm; his food, his sleep, his scramblings, his playthings, were all subjects of the deepest anxiety. Then Sir John, desiring to spare her "the sadness of farewell," perpetrated a moonlight flitting, which left her dismayed and desolate at the abrupt termination of the visit, and smarting with the intolerable consciousness of opportunities lost for saying what could now never be said. "All that passed," she said, "was like Sheridan's Chapter of Accidents." It was too much for her; she did not desire the repetition of a pleasure rated at a price higher than she could afford to pay. "I would not wish on any account," she told Lady Herschel in 1842, "to see either my nephew or you, my dear niece, again in this world, for I could not endure the pain of parting once more; but I trust I shall find and know you in the next."

She lived habitually in the past, and found the

present—as Mrs. Knipping, Dietrich's daughter said—
" not only strange, but annoying." Sometimes she
would rouse herself from a " melancholy lethargy " to
spend a few moments " in looking over my store of
astronomical and other memorandums of upwards of
fifty years' collecting, and destroying all that might
produce nonsense when coming through the hands of
a Block-kopf into the Zeitungen." Again she would
dip back into the career of the " forty-foot," or recall
the choral performance to which the tube had
resounded not far from sixty years before, "when I was
one of the nimblest and foremost to get in and out of
it. But now—lack-a-day—I can hardly cross the
room without help. But what of that ? Dorcas, in
the *Beggars' Opera*, says :

> "' One cannot eat one's cake and have it too ! ' "

That venerable instrument marked for her the *ne plus
ultra* of optical achievement. She would not admit the
sacrilegious thought of its being outdone. " I believe
I have water on my brains," she informed her nephew
in August, 1842, " and all my bones ache so that I can
hardly crawl ; and, besides, sometimes a whole week
passes without anybody coming near me, till they
stumble on a paragraph in the newspaper about
Gruithuisen's discoveries, or Lord Queenstown's great
telescope, which *shall* beat Sir William Herschel's all
to nothing ; and such a visit sometimes makes me
merry for a whole day."

From time to time she wrote books of " Recollections," which she forwarded, with anxious care, to
England. They contain nearly all that is intimately
known of Sir William Herschel's life. The entries in
her " Day-book " ceased finally only on September 3rd,

J

1845. In the hope of giving permanent form to the memories that haunted her, she began, at ninety-two, " a piece of work which I despair of finishing before my eyesight and life leave me in the lurch. You will, perhaps, wonder what such a thing can be as I may pretend to do; but I cannot help it, and shall not rest till I have wrote the history of the Herschels." " You remember," she added, " you take the work in whatever state I may leave it, and make the best of it at your leisure." It remained a piquant fragment. The fervour of her start was soon quenched by physical collapse, and she acknowledged her powerlessness " to do anything beside keeping herself alive." Her last letter to Collingwood was finished with difficulty, December 3rd, 1846. Monthly reports of her state, however, continued to be sent thither by Miss Beckedorff, who, with Mrs. Knipping, cared for her to the last.

In honour of her ninety-sixth birthday, the King of Prussia sent her, through Humboldt's friendly hands, the Gold Medal of Science; and on the following anniversary, March 16th, 1847, she entertained the Crown Prince and Princess for two hours. Not only with conversation; she sang to them, too, a composition of Sir William's, " Suppose we sing a Catch." She had a new gown and smart cap for the occasion ; and seemed " more revived than exhausted " by her efforts. Her last message to her nephew and his family—sent March 31st—was to say, with her "best love" " that she often wished to be with them, often felt alone, did not quite like old age with its weaknesses and infirmities, but that she, too, sometimes laughed at the world, liked her meals, and was satisfied with Betty's services."

On the 9th of January, 1848, she tranquilly breathed her last, and " the unquiet heart was at rest." She was buried beside her parents in the churchyard of the Gartengemeinde, at Hanover, with an epitaph of her own composition.* It records that the eyes closed in death had in life been turned towards the " starry heavens," as her discoveries of comets, and her participation in her brother's " immortal labours," bear witness to future ages. By her special request a lock of " her revered brother's " hair, and an old almanac used by her father, were placed in her coffin, which was escorted to the grave by royal carriages, and covered with wreaths of laurel and cypress from the royal gardens at Herrenhausen.

Caroline Herschel was not a woman of genius. Her mind was sound and vigorous, rather than brilliant. No abstract enthusiasm inspired her; no line of inquiry attracted her; she seems to have remained ignorant even of the subsequent history of her own comets. She prized them as trophies, but not unduly. The assignment of property in comets reminded her, she humorously remarked, when in her ninety-third year, of the children's game, " He who first cries ' Kick!' shall have the apple." Yet her faculties were of no common order, and they were rendered serviceable by moral strength and absolute devotedness. Her persistence was indomitable, her zeal was tempered by good sense; her endurance, courage, docility, and self-forgetfulness went to the limits of what is possible to human nature. With her readiness of hand and eye,

---

* " Der Blick der Verklärten war hienieden dem gestirnten Himmel zugewandt; die eigenen Cometen-Entdeckungen, und die Theilnahme an den unsterblichen Arbeiten ihres Bruders, Wilhelm Herschel, zeugen davon bis in die späteste Nachwelt."

her precision, her rapidity, her prompt obedience to a word or glance, she realised the ideal of what an assistant should be.

Herself and her performances she held in small esteem. Compliments and honours had no inflating effect upon her. Indeed, she deprecated them, lest they should tend to diminish her brother's glory. "Saying too much of what I have done," she wrote in 1826, "is saying too little of him, for he did all. I was a mere tool which *he* had the trouble of sharpening and adapting for the purpose he wanted it, for lack of a better. A little praise is very comforting, and I feel confident of having deserved it for my patience and perseverance, but none for great abilities or knowledge." Again: "I did nothing for my brother but what a well-trained puppy-dog would have done; that is to say, I did what he commanded me." And her entire and touching humility appears concentrated in the following sentence from a letter to her nephew: "My only reason for saying so much of myself is to show with what miserable assistance your father made shift to obtain the means of exploring the heavens."

The aim in life of this admirable woman was not to become learned or famous, but to make herself useful. Her function was, in her own unvarying opinion, a strictly secondary one. She had no ambition. Distinctions came to her unsought and incidentally. She was accordingly content with the slight and fragmentary supply of knowledge sufficing for the accurate performance of her daily tasks. No inner craving tormented her into amplifying it. The following of any such impulse would probably have impaired, rather than improved, her efficiency. The turn of her mind was, above all things, practical. She used formulæ as other women use pins, needles, and scissors, for certain definite purposes, but with complete indifference as to the mode of their manufacture. What was required of her, however, she accomplished superlatively well, and this was the summit of her desires. She shines, and will continue to shine, by the reflected light that she loved.

CHAPTER VII.

"The little boy is entertaining, comical, and promising," Dr. Burney wrote after his visit to Slough in 1797. John Frederick William Herschel was then five years old, having been born "within the shadow of the great telescope" March 7, 1792. He was an industrious little fellow, especially in doing mischief. "When one day I was sitting beside him," his aunt relates, "listening to his prattle, my attention was drawn by his hammering to see what he might be about, and I found that it was the continuation of many days' labour, and that the ground about the corner of the house was undermined, the corner-stone entirely away, and he was hard at work going on with the next. I gave the alarm, and old John Wiltshire, a favourite carpenter, came running, crying out, 'God bless the boy, if he is not going to pull the house down!'" And she wrote to him at Feldhausen; "I see you now in idea, running about in petticoats among your father's carpenters, working with little tools of your own, and John Wiltshire crying out, 'Dang the boy, if he can't drive in a nail as well as I can!'"

"John and I," she told his wife, "were the most affectionate friends, and many a half or whole holiday spent with me was dedicated to making experiments in chemistry, in which generally all boxes, tops of tea-canisters, pepper-boxes, teacups, etc., served for

SIR JOHN FREDERICK WILLIAM HERSCHEL, BART.

(*From a portrait painted by Pickersgill for St. John's College, Cambridge.*)

the necessary vessels, and the sand-tub furnished the matter to be analysed. I only had to take care to exclude water, which would have produced havoc on my carpet."

From a preparatory school kept by Dr. Gretton at Hitcham, he was sent, a delicate, blue-eyed lad, to Eton. His mother, however, happening to see him maltreated by a stronger boy, brought him home after a few months, and his education was continued by a Scotch mathematician named Rogers, a man of considerable ability. His pupil held him in high respect; yet, though he learned Euclid accurately from him, he told Dr. Pritchard afterwards that "he knew no more of its real bearing and intention than he knew of the man in the moon." The results of the home tuition were, none the less, exceedingly brilliant.

Herschel entered St. John's College, Cambridge, at the age of seventeen, and his aunt noted in her Diary that, from the time of his admittance to the University until he quitted it, he gained all the first prizes without exception. He graduated as Senior Wrangler and First Smith's Prizeman in 1813, a year in which honours were not cheap. Peacock, subsequently Dean of Ely, took second place, Fearon Fallows, the first Royal Astronomer at the Cape of Good Hope, came third, and Babbage withdrew from the competition, judging himself unable to beat, and not caring to be beaten by Herschel. Rivalry did not disturb their friendship. Having entered, together with Peacock, into a juvenile compact to do what in them lay " to leave the world wiser than they found it," they, in 1812, set about fulfilling it by the establishment of the " Analytical Society of Cambridge."

Its object was to substitute in England for Newton's fluxional method the more flexible and powerful calculus in use on the Continent; or, as Babbage expressed it, punning on the required change of notation, to uphold the principles of pure *D*-ism in opposition to the *Dot*-age of the University." The trio of innovators were full of enthusiasm, and they carried through a reform vital to the progress of British science. Herschel laboured zealously in the cause. In combination with his two allies, he translated Lacroix's elementary treatise on the Differential Calculus, which became a text-book at Cambridge ; and published, in 1820, an admirable volume of "Examples." "In a very few years," to use Babbage's words, "the change from dots to d's was accomplished; and thus at last the English cultivators of mathematical science, untrammelled by a limited and imperfect system of signs, entered on equal terms into competition with their Continental rivals." Herschel, writing in the *Quarterly Review*, playfully described the process by which this was brought about. "The brows of many a Cambridge moderator," he said, "were elevated, half in ire, half in admiration, at the unusual answers which began to appear in examination-papers. Even moderators are not made of impenetrable stuff; their souls were touched, though fenced with seven-fold Jacquier, and tough bull-hide of Vince and Wood. They were carried away with the stream, in short, or replaced by successors full of their newly acquired powers. The modern analysis was adopted in its largest extent."

John Herschel was one of Babbage's "chief and choicest companions," who breakfasted with him every Sunday after chapel, and discussed, during three or

four delightful hours, "all knowable, and many un-
knowable things." His life-long friendship with
Whewell began after his election to a Fellowship of
his College. It lent charm to the occasional residences
at Cambridge, which terminated in 1816, on his
attaining the dignity of Master of Arts. He celebrated
his coming of age at home, and was with his father at
Brighton when Campbell characterised him as "a
prodigy in science, and fond of poetry, but very
unassuming." His first publication was a paper on
" Cotes's Theorem," sent, in October, 1812, to the Royal
Society, of which body he was chosen a member,
May 27, 1813. This was followed by a series of
memoirs on various points of analysis, their signal
merit being recognised, in 1821, by the bestowal of
the Copley Medal. His investigations in pure
mathematics were carried no further; but he had
done enough to show his power and originality,
and materially to widen the scope of the new
methods.

He was in no hurry to choose a profession. Evenly
balanced inclinations demanded, circumstances indulged delay; so he paused. His father wished him
to enter the Church; but he preferred the law, and
was enrolled a student at Lincoln's Inn, January 24,
1814. The step was a simple formality. It committed
him to nothing. And, in fact, while nominally reading
for the Bar, his thoughts were running in a totally
different direction. Dr. Wollaston, whose acquaintance he made in London, fascinated him, and his
influence served to steady the helm of his intentions.
Having decided finally for a scientific career, he
returned to Slough, and plunged into experiments in
chemistry and physical optics.

On September 10th, 1816, he informed a correspondent that he was " going, under his father's direction, to take up star-gazing." This brief sentence gives the first tidings of an astronomical element in his life. Its growth was slow. He had no instinctive turn that way. It was through filial reverence that he resolved to tread in his father's footsteps. His self-denial received a magnificent reward. He took a place expressly reserved for him, as it might seem, beside his father as an explorer of the skies on the grandest scale. But for this moral purpose, he might have squandered time in a multiplicity of partial researches. So late as 1830 he told Sir William Rowan Hamilton : " I find it impossible to dwell for very long on one subject, and this renders my pursuit of any branch of science necessarily very desultory." His nebulæ and double stars saved him from being " everything by turns, and nothing long." Their collection and revisal, begun as a duty, grew to be irresistibly attractive, and John Herschel pledged himself definitively to astronomy.

His earliest undertaking was the re-examination of his father's double stars. Entered upon at Slough in 1816, it was continued from 1821 to 1823 at the observatory in Blackman Street, Southwark, of Mr., afterwards Sir James South, where, with two excellent refracting telescopes, of five and seven feet focal length, the colleagues measured 380 of Sir William Herschel's original pairs. Double stars want a great deal of looking after. Their discovery should be the prelude to long processes of investigation. It is of little interest unless diligently followed up. Each represents a system, individual in its peculiarities, and probably of most complex organisation. The

more such systems are studied, the more wonderful
they appear. Two associated stars have often proved,
on keener scrutiny, to be themselves very closely
double; and in other cases, disturbed motion has
revealed the existence of obscure masses—planets
on a colossal scale, possibly the spacious abodes of
unimaginable forms of life.

The "Astronomy of the Invisible," however, was
still in the future when Herschel and South did their
work. Facts relating to binary revolutions were
scantily forthcoming, and the science to be founded on
them had been rather indicated than established.
Fresh observations were then needed to ascertain how
the circling stars had behaved since 1802. The results
proved highly satisfactory. In Francis Baily's words,
" The remarkable phenomena first brought to light by
Sir William Herschel were abundantly confirmed, and
many new objects pointed out as worthy the attention
of future observers." To take a couple of examples.
Eta Coronæ was found to have described, since 1781,
one entire round, and to be just starting on a second.
Again, Tau Ophiuchi had been perceived, by the elder
Herschel, at his first sight of it in April, 1783, to be
" elongated." " One half of the small star," he said, "if
not three-quarters, seems to be behind the large star."
This effect was imperceptible to his son. It had
become entirely effaced in the course of forty years.
The star was, in 1823, perfectly round; it had, as it
were, absorbed its companion. By slow degrees, however, the two came into separate view, and now form
an easy telescopic object. Their period of revolution
is not less than two centuries. Another point of
special interest was the detection of marked eccen-
tricity in a stellar orbit—that of Xi Ursæ Majoris.

These stars perform their circuits in just sixty years; but in 1821 their apparent speed was so great that changes in their relative positions could be determined from month to month. For these observations, published with notes and discussions in the *Philosophical Transactions* for 1824, Herschel and South received the Lalande Prize of the French Academy in 1825, and the Gold Medal of the Astronomical Society in 1826. In the latter distinction, Wilhelm Struve and Amici of Modena were associated with them. These four were the only double star observers then living.

Their exertions served to define more closely the circumstances of stellar movement. The crucial question could now be put, whether they are governed by the force that binds the planets to the sun, or by some other form of attractive influence. In other words, is the law of gravitation universal? An answer could only be obtained experimentally, by computing, on gravitational principles, the paths of the best-known pairs, and then *trying the fit*. If the stars, as time went on, kept near their predicted places, the unity of nature in this respect might be safely inferred; although considerable discrepancies might in any case be expected, owing to errors of measurement minute in themselves, but large relatively to curves reduced by distance to hair-breadth dimensions.

This kind of inquiry was fairly started in 1827, when Savary computed the orbit of Xi Ursæ. His success made it almost certain that the pair moved under the planetary regimen, conformed to, there is no reason to doubt, by all binaries. John Herschel, although not the first, was the most effective early investigator of stellar orbits. His method, described before the Royal Astronomical Society January 13, 1832, and approved by the award of its Gold Medal in 1833, went to the root of the matter. The author declared it a mere waste of time to attempt to deal, by any refined or intricate process of calculation, with data so uncertain and irregular as those at hand. "Uncertain and irregular," it must be repeated, because referred to a scale on which tenths of a second assume large proportions. He accordingly discarded, as mere pedantic trifling, such analytical formulæ as those employed by Savary and Encke, and had recourse to a graphical process, in which "the aid of the eye and hand" was used to "guide the judgment in a case where judgment only, and not calculation, could be of any avail." The operation which he went on to explain was commended by Sir George Airy for its "elegance and practical utility." Nothing more appropriate could have been devised than this plan, at once simple, ingenious, and accommodating, for drawing a curve representative of the successive relative positions of double stars. Its invention effectively promoted acquaintance with their orbits; most of those at present known having, indeed, been calculated with its aid.

In 1821, Herschel travelled, in Babbage's company, through Switzerland and Italy. His only recorded adventure was an ascent of Monte Rosa. In the following year he visited Holland with James Grahame, the learned author of a "History of America"; and on the removal of South's observatory to Passy, he again went abroad, starting with Babbage, but returning alone. This time he made a

number of scientific acquaintances. His father's
name worked like a spell. "I find myself," he said,
"for his sake, received by all men of science with
open arms." His modesty forbade him to remember
that his own merits were already conspicuous. In
Paris, Arago and Fourier showed him all possible
attentions; he was welcomed at Turin "like a
brother" by Plana, "one of the most eminent mathe-
maticians of the age;" at Modena, Amici was, if
possible, still more cordial. "He is the only man,"
Herschel told his aunt, "who has, since my father,
bestowed great pains on the construction of specula."
"Among other of your inquiring friends," he continued,
"I should not omit the Abbé Piazzi, whom I found ill
in bed at Palermo, and who is a fine, respectable old
man, though, I am afraid, not much longer for this
world. He remembered you personally, having himself visited Slough."

On July 3 Herschel "made the ascent of Etna,
without particular difficulty, though with excessive
fatigue." On the summit, reached before sunrise, by
"a desperate scramble up a cone of lava and ashes,
one thousand feet high," he found himself "enveloped
in suffocating sulphurous vapours"; and "was glad
enough to get down," after having made a reading
of the barometer in concert with the simultaneous
observations of the brothers Gemellaro at Catania
and Nicolosi. The same night he arrived at Catania
"almost dead" from the morning's arduous climb,
"and the dreadful descent of nearly thirty miles,
where the mules could scarce keep their feet."

In traversing Germany, he deviated to Erlangen,
where Pfaff was engaged in translating Sir William
Herschel's writings; and visited Encke, Lindenau,

and Harding, at Seeberg, Gotha, and Göttingen. With Göttingen he had a special tie through his creation, in 1816, an honorary member of the University; and at Göttingen, too, he hoped to meet Gauss—a man of strange, and—to the lay mind—unintelligible powers. "Gauss was a god," one of his fellow-mathematicians said of him; but the "god" was on this occasion absent—feasting with the "blameless Ethiopians," perhaps, like the Homeric deities when wanted. He was reported "inconsolable" for the lost opportunity, which seems never to have recurred.

From Munich Herschel wrote to his aunt, in view of his approaching visit to Hanover:—"I hope you haven't forgotten your English, as I find myself not quite so fluent in this language (German) as I expected. In fact, since leaving Italy, I have so begarbled my German with Italian that it is unintelligible both to myself and to everyone that hears it: and what is very perverse, though when in Italy I could hardly talk Italian fit to be heard, I can now talk nothing else, and whenever I want a German word, pop comes the Italian one in its place. I made the waiter to-day stare (he being a Frenchman) by calling to him, 'Wollen Sie avere la bontà den acet zu apportaren!' But this, I hope, will soon wear off."

His next foreign holiday was spent in France. He had designed a new instrument for measuring the intensity of the sun's radiations, and was eager to experiment with it alternately at high and low levels, for the purpose of determining the proportion of solar heat absorbed by the earth's atmosphere. This method was employed with fine effect by Professor

Langley on Mount Whitney in 1881. Herschel carried his "actinometer" to the top of the Puy de Dôme in September 1826, and waited at Montpellier for "one day of intense sunshine," in order to procure his second term of comparison. The Puy de Dôme, with its associated three hundred summits, strongly allured him. "I have been rambling over the volcanoes of Auvergne," he wrote from Montpellier, September 17, "and propose, before I quit this, to visit an extinct crater which has given off two streams of lava at Agde, a town about thirty miles south of this place on the road to the Spanish frontier. Into Spain, however, I do not mean to go—having no wish to have my throat cut. I am told that a regular diligence runs between this and Madrid, and is as regularly stopped and robbed on the way."

This exploratory turn alarmed Miss Herschel. "I fear," she replied, "you must often be exposed to great dangers by creeping about in holes and corners among craters of volcanoes." He was, nevertheless, only dissuaded by his mother's anxious remonstrances from pursuing their study in Madeira and Teneriffe.

In the autumn of 1827, Babbage accompanied him to Ireland. The young Astronomer Royal, Sir W. R. Hamilton, was unluckily absent at the time of their visit; but he sent Herschel, by way of compensation, one of his brilliant optical essays, and a correspondence sprang up from which a lasting friendship developed.

Herschel's scientific occupations at home were meanwhile various and pressing. He co-operated in the foundation of the Astronomical Society, and became in 1821 its first foreign secretary. In 1824 he undertook the more onerous duties of secretary to the Royal Society, and rented a house in Devonshire Street for the three years of his term of office. Astronomy, it might have been feared, should be at least temporarily shelved; yet he informed his aunt, April 18, 1825, "A week ago I had the twenty-foot directed on the nebulæ in Virgo, and determined the right ascensions and polar distances of thirty-six of them. These curious objects I shall now take into my especial charge—nobody else can see them."

His telescope, in fact, then held the championship. It was constructed in 1820 by himself, under his father's directions, on the "front view" plan, the speculum being eighteen inches in diameter, and of twenty feet focal length. With it he executed, in 1824, a fine drawing of the Orion Nebula, with which "inexplicable phenomenon" he was profoundly impressed. It suggested to him no idea of a starry composition, and he likened its aspect to that presented by the "breaking up of a mackerel sky, when the clouds of which it consists begin to assume a cirrous appearance."

In July, 1828, he succeeded in discerning the two Uranian satellites, Oberon and Titania, *authentically* discovered by his father. They had not been seen, except incidentally at Slough, for thirty years. His pursuit of them, continued at intervals until 1832, had the result of confirming, while slightly correcting, Sir William Herschel's elements of their motions. On September 23, 1832, he perceived Biela's comet as a round, hazy object without a tail. It closely simulated a pretty large nebula. A small knot of very faint stars lay directly in its path, and, having before long overtaken them, it "presented, when on the

K

cluster, the appearance of a nebula partly resolved into stars, the stars of the cluster being visible through the comet." They shone undimmed, he estimated, from behind a veil of cometary matter 50,000 miles thick. Yet, only a month later, the remote prospect of a collision with this tenuous body threw Europe into a panic.

After Sir William Herschel's death, his son formed the project of collecting into a memorial volume all his published papers; but he decided before long that he could add more to his fame by pursuing and verifying his observations than by reprinting them. The keynote of his life's activity was struck in these words. His review of the 2,500 Herschelian nebulæ, more than half of which were invisible with any instrument except his own, was begun in the summer of 1825, and terminated in 1833. The assiduity with which it was prosecuted appeared by its completion in little more than half the time judged necessary for the purpose by the original discoverer. Yet he was not exempt from discouragement. "Two stars last night," he wrote, July 23, 1830, "and sat up till two waiting for them. Ditto the night before. Sick of star-gazing—mean to break the telescopes and melt the mirrors." Very few glimpses of this seamy side to the occupation are afforded us by either of the Slough observers. Modern astronomers, by comparison, would seem, like the Scotchman's barometer, to have "lost all control over the weather."

The efficacious promptitude with which John Herschel swept the skies appears truly wonderful when we remember that he was without a skilled assistant. No ready pen was at hand to record what he saw, and how he saw it; he was, by necessity, his own amanuensis; and writing by lamplight unfits the eye for receiving delicate impressions. Yet a multitude of the objects for which quest was being made were of the last degree of faintness. The results were none the less admirable. Embodied in a catalogue of 2,307 nebulæ, of which 525 were new, they were presented to the Royal Society July 1, 1833, and printed in the *Philosophical Transactions* (vol. cxxiii.). Annotations of great interest, and over one hundred beautiful drawings, enhanced the value of the memoir.

Herschel was struck, in the course of his review, by the nebulous relations of double stars. A close, faint pair at the exact centre of a small round nebula in Leo; stellar foci in nebular ellipses; and a strange little group consisting of a trio of equidistant stars relieved against a nebulous shield, were specimen-instances illustrating "a point of curious and high physical interest."

He also drew attention to "the frequent and close proximity to planetary nebulæ of minute stars which suggest the idea of accompanying satellites. Such they may possibly be." If so, their revolutions might eventually be ascertained; and he urged the desirability of exact and persistent determinations of the positions of these satellite-stars. "I regret," he concluded, "not having sufficiently attended to this in my observations, the few measures given being hurried, imperfect, and discordant." Up to the present, these supposed systems have remained sensibly fixed; but they have been a good deal neglected. Mr. Burnham's observations, however, with the Lick refractor in 1890–1, may supply a basis for the future detection

K 2

of their movements in periods probably to be reckoned by millenniums.

The orbital circulation of compound nebulæ must be at least equally slow. They are most diverse in form and arrangement. " All the varieties of double stars as to distance, position, and relative brightness," Herschel wrote, " have their counterparts in nebulæ; besides which, the varieties of form and gradation of light in the latter afford room for combinations peculiar to this class of objects." Such, for instance, as the disparate union of an immensely long nebulous ray in Canes Venatici with a dim round companion, a small intermediate star occupying possibly the centre of gravity of the system.

Herschel's drawings of double nebulæ have gained significance through their discussion, in 1892, by Dr. T. J. J. See of Chicago. They are now perceived to form a series aptly illustrative of the process, theoretically investigated by Poincaré and Darwin, by which a cooling and contracting body, under the stress of its consequently accelerated rotation, divides into two. If it be homogeneous in composition, its " fission" gives rise to two equal masses, presumed to condense eventually into a pair of equal stars. Disparity, on the other hand, between the products of fission indicates original heterogeneity; so that a large nebula must be of denser consistence than a smaller one physically connected with it. The chemical dissimilarity of the stars developed from them might explain the colour-contrasts often presented by unequal stellar couples. This view as to the origin of double nebulæ, and through them of double stars, although doubtless representing only a fragment of the truth, gives wonderful coherence to Herschel's faithful delineations of what his telescope showed him.

No one before him had completely seen the " Dumb-bell " nebula in Vulpecula. Sir William Herschel had perceived the " double-headed shot " part of this "most amazing object," but had missed the hazy sheath which his successor noticed as filling in the elliptic outline. He perceived similarly (unaware of Schröter's observation) that the interior of the Ring-nebula in Lyra is not entirely dark; and compared the effect to that of fine gauze stretched over a hoop. An exceedingly long, nebular ellipse in Andromeda, with a narrow interior vacuity, left him " hardly a doubt of its being a thin, flat ring of enormous dimensions, seen very obliquely." A photograph taken by Dr. Roberts, in 1891, corresponds strikingly with Herschel's drawing. Some specimens of " rifted nebulæ," were also included in the collection of 1833. They are double or even triple parallel rays, fragments, apparently, of single primitive formations. Herschel might well assert that " some of the most remarkable peculiarities of nebulæ had escaped every former observer."

Both by the Royal, and by the Royal Astronomical Societies, medals were, in 1836, adjudged to this fine work. Its progress was accompanied by the discovery of 3,347 double stars, as well as by the re-measurement of a large number of pairs already known. The whole were drawn up into eight catalogues, presented at intervals to the Astronomical Society, and printed in their Memoirs. A good many of them would, nevertheless, be rejected by modern astronomers as "not worth powder and shot," the stars composing them being too far apart to give more than an infinitesimal chance of mutual connection. From May 1828 onwards, these measures were made with "South's *ci-devant* great equatorial," purchased by Herschel. The object-glass, by Tulley, was five inches in diameter. With a twelve-inch refractor, its successor in South's observatory on Campden Hill, Herschel detected, on its trial-night, February 13, 1830, the sixth star in the "trapezium" of Orion. This minute object was then about one-third as bright as the fifth star in the same group, discovered by Robert Hooke in 1664, but forgotten, and re-discovered by Struve in 1826. A slow gain of light in Herschel's star is not improbable.

He refused, in 1826, to compete for the Lucasian Professorship of Mathematics at Cambridge. It was practically at his disposal, since all agreed that no one could better than Herschel have filled the chair once occupied by Newton. He was, however, disinclined for an University career, and had undertaken labours incompatible with it. In 1830 he stood as the "scientific candidate" for the presidentship of the Royal Society, against the Duke of Sussex. His defeat was by "a ridiculously small majority." "I had no personal interest in the contest," he wrote to Sir William Hamilton. "Had my private wishes and sense of individual advantage weighed with me in opposition to what, under the circumstances, was an imperative duty, I should have persisted in my refusal to be brought forward; but there are situations where one *has* no choice, and such was mine."

He made Hamilton's personal acquaintance at a dinner of notabilities, given by the Duke of Sussex, in March, 1832. An invitation to Slough followed, and

Hamilton, arriving " in a beautiful star-time," enjoyed celestial sights that seemed the opening of a new firmament.

Herschel married, March 3, 1829, Margaret Brodie, second daughter of the Rev. Alexander Stewart, of Dingwall, in Ross-shire. The event—not merely by convention a " happy " one—gave great satisfaction to his numerous friends. Miss Herschel was beside herself with glad emotion. " I have spent four days," she informed him on his wedding-day, "in vain endeavours to gain composure enough to give you an idea of the joyful sensation caused by the news. But I can at this moment find no words which would better express my happiness than those of Simeon : " Lord, now lettest thou thy servant depart in peace." But there was no finality in her desires for this brilliant scion of her race. His domestic felicity did not long content her ; she craved worldly distinctions. When, after the accession of William IV., a shower of honours was let fall, she began to think plain " John Herschel, Esq.," an address very inadequate to his merits. " Dr. Grosskopf," the husband of one of her nieces, "has been *zum Ritter ernannt* by his present Majesty," she wrote discontentedly. "So was Dr. Mükry last week. If all is betitled in England and Germany, why is not my nephew, J. H., a lord or a wycount (*sic*) at least ? General Komarzewsky used to say to your father, ' Why does not King George III. make you Duke of Slough ? ' "

An instalment of her wishes was granted by his creation, in 1831, a Knight of the Royal Hanoverian Guelphic Order ; and she lived to see him a baronet. She had no inkling of his approaching journey to the Cape when he came to see her in June, 1832, although the visit was designed as a farewell. Hanover itself, too, had for him an ancestral charm.

"It was only this evening," he wrote home, "that, escaping from a party at Mrs. Beckedorff's, I was able to indulge in what my soul has been yearning for ever since I came here—a solitary ramble out of town, among the meadows which border the Leine-strom, from which the old, tall, sombre-looking Marktthurm, and the three beautiful lanthorn steeples of Hanover are seen as in the little picture I have often looked at with a sort of mysterious wonder when a boy, as that strange place in foreign parts that my father and uncle used to talk so much about, and so familiarly. The *likeness* is correct, and I soon found the point of view."

Almost from the beginning of his surveying operations, Herschel cherished the hope of extending them to the southern hemisphere. But during his mother's lifetime, he took no steps towards its realisation. The separation would have been cruel. Her death, however, on January 6th, 1832, at the age of eighty-one, removed this obstacle, and the scheme rapidly took shape. The station originally thought of was Paramatta, in New South Wales; but Dunlop's observations there anticipated him, and he reflected with disappointment that "the cream of the southern hemisphere had been skimmed" before his turn came. He learned afterwards that nothing important in the "sweeping" line had been done at Paramatta; he had virgin skies to explore. A trip to the Himalayas was his next ambition; and one of the recommendations of the Cape of Good Hope was its being "within striking distance of India." But to India he never went. The Cape was beyond question the most suitable locality for his purpose, and it would have been waste of time to have left it, even temporarily, for any other. He was offered a free passage thither in a ship of war, but preferred to keep his enterprise altogether on a private footing. So, having embarked with his wife, three children, and instrumental outfit, on board the *Mountstuart Elphinstone*, he left the shores of England, November 13, 1833.

CHAPTER VIII.

EXPEDITION TO THE CAPE.

THE voyage was prosperous, but long. Nine weeks and two days passed before the welcome cry of "Land" was heard; and it was in the dawn of January 15, 1835, that Table Mountain at last stood full in view, with all its attendant range down to the farthest point of South Africa," outlined, ghost-like, in clear blue. The disembarkation of the instruments and luggage took several days. They filled fifteen large boats, and a single onslaught of the south-easterly gale, by which at that time of the year Cape Town is harried, might easily have marred the projected campaign. All, however, went well.

The travellers were welcomed by Dr. Stewart, one of Lady Herschel's brothers, and enthusiastically greeted by the Royal Astronomer, Sir Thomas Maclear. They made no delay in fixing their headquarters.

"For the last two or three days," Herschel wrote to his aunt, January 21, "we have been looking for houses, and have all but agreed for one, a most beautiful place four or five miles out of town, called 'The Grove.' In point of situation it is a perfect paradise in rich and magnificent mountain scenery, and sheltered from all winds, even the fierce south-easter, by thick surrounding woods. I must reserve for my next all description of the gorgeous display of flowers which adorns this splendid country, as well as the astonishing brilliancy of the constellations."

"The Grove" resumed its old Dutch name of "Feldhausen" during Herschel's occupation of it; and as "Feldhausen" it will always be memorable in astronomical history as the scene of the first effective exploration of the southern heavens. The place is essentially unchanged. Only an avenue of fir-trees has been planted by way of approach to the house, a solid Dutch structure, with a disconsolate-looking garden in front; while in an adjacent field, carpeted with yellow lupins every spring, and redolent of their perfume, an obelisk has been erected on the former site of the great reflector. Above, to the west, towers the *gable-end* of Table Mountain, and an exuberant growth of oaks and pines softens the sternness of its "mural precipices."

The neighbourhood was, in those days, lonely in the human sense, although otherwise over- and ill-populated. Wolves and jackals abounded in the forests; venomous snakes slid through the grass; baboons had the run of the country; even the lion and the hippopotamus were scarcely yet extinct in the Cape Peninsula. Many a wild hyæna-shriek startled the astronomer at his nightly toil; and Dr. Whewell reported that he had "spent one night in tiger-hunting, but seemed to think it poor sport compared with the *chasse aux étoiles doubles.*" *Tiger*, it should be explained, is a local name for a species of leopard: no true tigers have ever been encountered in Africa.

His twenty-foot began its activity February 22nd, and the refractor, which was equatorially mounted in a revolving dome, was ready early in June. "But I am sorry to say," he told Miss Herschel, "that the nights in which it can be used to advantage are rare." And he lamented to his brother-in-law and intimate friend, Mr. James C. Stewart, that, during the hot season, " the stars tremble, swell, and waver most formidably." The Cape heavens are indeed often exasperating. On nights meteorologically quite fine, the dismayed astronomer not uncommonly sees the stars " walking about " in the field of view; and a mere handful of cloud will, at other times, with incredible swiftness, spread over the whole face of the sky. Still, compensation is, sooner or later, sure to come in a run of magnificent observing weather. This was Herschel's experience. He informed Francis Baily, October 23rd, 1834, that " the definition was far beyond anything experienced in England." After rain especially, superb opportunities were afforded, when

"The starry sequence of nocturnal hours" *

might be unbroken, perhaps for a week together, by a single adverse incident of climate.

Herschel took three specula with him to the Cape; one made by his father, another by himself with his father's aid, and a third, of his own exclusive manufacture. Their rapid tarnishing kept them in constant circulation from the tube to the polisher. After half a dozen nights they had lost all brilliancy; at the end of three months, they were more than purblind. He acquired, however, such facility and skill in the use of his polishing machine, that he was able, in 1835, to report his mirrors as " more perfect than at any former time."

He made astonishingly quick progress in observation. On October 24th, 1835, Miss Herschel was informed, "I have now very nearly gone over the

* R. Garnett, "Iphigenia in Delphi."

whole southern heavens, and over much of it often.
In short, I have, to use a homely phrase, broken the
neck of the work, and my main object now is to
secure and perfect what is done."

His sweeps yielded a harvest of 1,202 double stars,
and 1,708 nebulæ and clusters, only 439 of which had
been previously registered. Among the novelties were
a faint, delicate miniature of the ring-nebula in Lyra,
and five planetaries. One of these he described as "of
a beautiful greenish-blue colour, a full and intense
tint." This lovely object, situated in Centaur, is sometimes distinguished as "*the* blue planetary"; although
its hue is shared by all the members of its class. The
nature of their spectrum, in fact, obliges them to be
more or less green.

Sir John Herschel applied the term "falcated" to
two curious nebulæ belonging, undoubtedly, to the
later recognised "spiral" class. He perceived besides
in oval nebulæ the annular lines of structure emphasised in Dr. Roberts's photographs. He remarked,
further, that "as the condensation increases toward
the middle, the ellipticity of the strata diminishes."

His study of the Magellanic Clouds gave the first
idea of their composition. He showed them to be
aggregations on a vast scale of every variety of cosmical product. "When examined through powerful
telescopes, the constitution of the Nubeculæ, and
especially of the Nubecula Major, is found to be of
astonishing complexity." He drew up a preliminary
catalogue of 1,163 stars, nebulæ, and clusters included
in them, the conjunction of which was really decisive as
to nebular status. For he showed, from the elementary
principles of trigonometry, that, taking the Greater
Cloud to be roughly spherical in shape, its nearest and

remotest parts could differ in distance from ourselves
by little more than one-tenth the distance of its centre.
The fact was thus demonstrated that seventh and
eighth-magnitude stars and irresolvable nebulæ co-
exist within those limits.   He stopped short, however, of the conclusion drawn by Whewell and Spencer,
that the stellar and nebular sub-kingdoms are not
only locally intermixed, but inseparably united.

The Magellanic Clouds are the most conspicuous
features of the barren south polar heavens.   Round
the Lesser Cloud especially, the sky, Herschel said,
"is most oppressively desolate."   And again : "The
access to the Nubecula Minor on all sides is through a
desert."   One of the separate inmates of the Larger
Cloud is the "great looped nebula," compared by
Herschel to "an assemblage of loops," the complicated
windings of which make it "one of the most extraordinary objects which the heavens present."   To
the eye of the present writer it resembled a shining
strip of cellular tissue hung up against the sky.
The "lace-work nebula" in Cygnus is of the same
type ; but here the tracery of nebula is closely
followed by a tracery of stars.   Truly, "A most
wonderful phenomenon !" as Herschel exclaimed in
contemplating it.

The first photographs of the Magellanic Clouds
were taken in 1890–91 by Mr. Russell of Sydney.
They contained an extraordinary revelation.   Both
objects came out in them as gigantic spirals.   Their
miscellaneous contents are then arranged according
to the dictates of a prevalent, though unexplained
cosmical law.   The Nubecula Major is a double
vortex, and the extent of its outlying portions, invisible except to the camera, is at least eight times

that of the central mass; but they conform to the same helical lines.

Herschel catalogued 1,203 stars strewn over the surface of the famous Argo nebula, and devoted several months to its delineation. This he found "a work of great labour and difficulty." While at the telescope he often half surrendered to despair "of ever being able to transfer to paper, with even tolerable correctness, its endless details." "Language cannot easily convey," he said, "a full impression of the beauty and sublimity of the spectacle this nebula offers when viewed in a sweep, ushered in by so glorious and innumerable a procession of stars, to which it forms a sort of climax." Only the Orion nebula may be thought to surpass it in "magnitude, complexity, and brightness." Its most characteristic feature is an abrupt vacuity, of a "lemniscate oval" shape, from which it derives the name of the "Key-hole Nebula." The value of Herschel's drawing of this grand object has been accentuated by its photographic portrayal. Their comparison betrays, in fact, the occurrence in the interval of what appears to be a vast change. Already, in 1871, Mr. Russell missed with surprise a prominent feature in the Feldhausen picture; and its failure to appear on photographic plates exposed for eight hours, yet impressed with innumerable stars previously unseen, raised something more than a presumption that, as Mr. Russell said, "a well-defined and brilliant portion of this nebula vanished between 1837 and 1871." Its disappearance was independently verified by Dr. Gill, Royal Astronomer at the Cape. With a total exposure of more than twelve hours, in March, 1892, he secured a magnificent representation of this wonderful object, fundamentally agreeing with

Herschel's, save only as regards the mass of bright nebulosity vainly looked for by Mr. Russell. The "swan-shaped" or "trident-like" structure was clean gone! That is to say, the matter composing it had ceased to be luminous. It should be added that Mr. Ranyard, whose special experience lent weight to his opinion, thought it unsafe to trust much to comparisons of drawings of such baffling objects, either among themselves or with photographs.

Before leaving the Cape, Herschel witnessed an event testifying surprisingly to the *vitality* of this nebula. In a condensed tract close to the dark "keyhole," he was accustomed to see the bright star Eta Argûs. It gave no sign of being variable until, on December 16, 1837, he perceived with amazement that it had, all at once, nearly tripled in brightness. After this sudden leap, it mounted gradually to the level of Alpha Centauri, then slowly declined. It just matched Aldebaran when Herschel lost sight of it in March, 1838. A second, and even more vigorous outburst was watched by Sir Thomas Maclear in 1843. It then overtopped every star except Sirius, and for seven subsequent years rivalled the splendour of Canopus. No notice was at first taken of its colour; but it was redder than Mars in 1850, and reddish it still remains, in its low estate of invisibility to the naked eye. But since bright lines of hydrogen show in its photographed spectrum, we may suspect that—

"Even in its ashes live its former fires,"

and that, consequently, its vicissitudes are not yet terminated. The instability of its character was virtually discovered at Feldhausen. Except by Burchell, the African traveller, no previous suspicion of it had been entertained; the numerous facts denoting that the star's past behaviour had been abnormal were collected by Sir John Herschel after it had been caught *in flagrante delicto*. In his belief, it had no physical connection with, but was merely projected upon, the nebula. But since then the nebular relations of blazing stars have been strongly underlined. The mass of circumstantial evidence now accumulated on the point fully warrants the assertion that Eta Argûs makes an integral part of the formation it once illuminated.

A cluster in the constellation of the Cross, unique in the varied and brilliant tints of its principal components, was compared by Herschel to "a gorgeous piece of fancy jewellery." Within the space of $\frac{1}{48}$th part of a square degree, he determined the places of no less than 110 of them, referred to Kappa Crucis, a rosy orb round which they are irregularly scattered. The colour-effects in this beautiful ornament of the sky need large apertures for their full display.

An object showing to the eye as a hazy star of the fourth magnitude was entitled by Bayer in 1603 Omega Centauri. Herschel's twenty-foot disclosed it as "a noble globular cluster, beyond all comparison the richest and largest in the heavens." Dr. Gill obtained an admirable photograph of it May 25, 1892. The stars composing it are literally countless. On a plate exposed for two hours at Arequipa, Mr. Solon I. Bailey reckoned nearly 6,400; yet he made no allowance for those "too faint and closely packed" to be perceptible except as a "mottled grey background between the distinct images."

Somewhat inferior to Omega Centauri in size, though not at all in beauty, is 47 Toucani. So

L

obvious is it to the naked eye that, for several nights
after his arrival in Peru, Humboldt took it for a comet.
Central condensation in this cluster appeared to
Herschel as if marked off into three distinct stages;
and to his delighted perception the whole interior
offered, by its roseate hue, an exquisite contrast to
the silvery radiance of the outer portions. No other
observer has, however, noticed this chromatic peculiarity. The structure of 47 Toucani is almost perfectly uniform. It is broken by none of the "dark
lanes," rifts, or tunnels which so curiously diversify
many globular clusters. The usual hirsute aspect
lent by the spreading abroad of *tentacles*, or radiating
stellar streams, is likewise scarcely distinguishable
either in 47 Toucani or Omega Centauri. Indeed,
Mr. Bailey noticed that the photographic images of
both were all but perfectly circular. In a future age
this may be otherwise. Streams of stars will, perhaps,
set outward from these grand assemblages, leaving
vacancies behind. Thus, if it be permissible to judge
of the relative antiquity of clusters by their advance
towards disruption, 47 Toucani and Omega Centauri
may be reckoned among the youngest of the globular
kind existing in the heavens.

The mechanism of clusters has received little attention from any astronomer beside Herschel. And a
solution of an ideal case of the problem it presented
was the utmost he could achieve.

"A quiescent spherical form," he wrote in 1833,
"may subsist as the bounding outline of an immense
number of equal stars, uniformly distributed through
its extent. In such a state of things each star might
describe an ellipse in any plane, and in any direction
in that plane, about the common centre without the

possibility of collision. If the form be not spherical, and the distribution of the stars not homogeneous, the dynamical relations become too complicated to be distinctly apprehended."

But the more closely these aggregations are examined, the less likely does it seem that they in any sense represent "quiescent forms." The arrangement of the stars composing them rather suggests their being outward bound into the ocean of surrounding space, although the orders that they carry are to us sealed.

Herschel subsequently altered his views regarding the composition of clusters, and threw out in 1847 "the possibility of masses of luminous matter—of whatever density or rarity, of whatever bulk or minuteness—forming a connected system, and being prevented from collapse or from mutual interference by the resistance of a transparent and non-luminous medium." For a "dynamical" he, in short, substituted a "statical equilibrium," the interposed medium lending unity to the mixed aggregate, and enabling it to rotate, as a whole, upon an axis. But the rotation is more than questionable. It seems to be precluded by the ragged contours and indeterminate boundaries of all starry collections. Photographic evidence, on the other hand, favours Sir John Herschel's surmise as to the composite nature of clusters. Some at least evidently unite within themselves the "two sidereal principles." The stellar points they mainly consist of are immersed in, or linked together by, shining nebulous stuff.

Herschel provided a southern sequel to his father's star-gauging work by counting 70,000 stars in 2,300 fields. Their distribution was in complete accordance

L 2

with the results of the earlier experiments. " Nothing can be more striking," Sir John wrote, " than the gradual, but rapid increase of density on either side of the Milky Way as we approach its course." The existence of an " ecliptic of the stars " (in Lambert's almost prophetic phrase) was demonstrated. Or, as Herschel himself put it, the plane of the Galaxy " is to sidereal, what the ecliptic is to planetary astronomy, a plane of ultimate reference, the ground-plan of the sidereal system." He estimated, from the basis of his gauge-reckonings, that his twenty-foot reflector was capable of showing, in both hemispheres, about five and a half million stars. The smallest of these would be of 14·5 magnitude, on the strict photometric scale. But, unless his valuation was greatly too small, there must be a conspicuous falling off in stellar density beyond the region of tenth or eleventh magnitude. If this be so, scarcely one-quarter of the expected stars will make their appearance on the plates of the International Survey.

The grand feature of southern celestial scenery is the splendour of the Milky Way. One of the galactic condensations in Sagittarius actually seems to start out from the sky in a definite globular form; and the darkness of the great rift beginning near the Cross is so intensified by contrast with the strongly luminous branches it separates, as to throw the blackness of the exterior heavens *into the shade*. This part of the Milky Way may even be seen in southern latitudes—as it was by the present writer—reflected from a glassy ocean-surface. The section passing from Centaur through the Ship to Orion is, in some respects, still more striking. Captain Jacob remarked at Madras that

" the general blaze from this portion of the sky is such as to render a person immediately aware of its having risen above the horizon, though he should not be at the time looking at the heavens." Herschel commented on the singular interruptions of the shining zone by obscure spaces in Scorpio, near Alpha Centauri, and elsewhere ; and admired the enhancement afforded to its magnificence by " a marvellous fringe of stars" attached pretty regularly to its southern border. " It is impossible," he wrote to Sir William Hamilton in June, 1836, " to resist the conviction that the Milky Way is not a stratum, but a ring."

His telescopic analysis disclosed in it a variety and complexity of structure for which he was wholly unprepared. " Great cirrous masses and streaks " of galactic light presented themselves in Sagittarius ; and, as the telescope moves, the appearance is that of clouds passing in a *scud*." " The Milky Way," he continued, " is like sand, not strewn evenly as with a sieve, but as if flung down by handfuls, and both hands at once, leaving dark intervals, and all consisting of stars of the lowest magnitudes," down to nebulosity, in a most astonishing manner." As he proceeded, the stars became " inconceivably numerous and minute. There must be millions on millions, and all most unequally massed together ; yet they nowhere run to nuclei, or clusters much brighter in the middle."

In some regions, the formation proved unfathomable ; all traces of stellar *texture* disappeared. In others, it was plainly perceived to consist of portions differing exceedingly in distance, but brought by projection into nearly the same visual line. Near the Trifid Nebula, " we see foreshortened," he said, " a vast and illimitable area scattered over with discontinuous masses and aggregates of stars, in the manner of the cumuli of a mackerel-sky, rather than of a stratum of regular thickness and homogeneous formation."

These varied observations compelled him to reject decisively Olbers's hypothesis of light-extinction in space. For, if the possible range of ethereal messages be restricted in one direction, it must be equally restricted in all. "We are not at liberty," he reasoned, "to argue that in one part of the circumference of the galaxy our view is limited by this sort of cosmical veil which extinguishes the smaller magnitudes, cuts off the nebulous light of distant masses, and closes our view in impenetrable darkness; while, at another, we are compelled, by the clearest evidence telescopes can afford, to believe that star-strewn spaces *lie open*, exhausting their powers and stretching out beyond their utmost reach." These objections seem fatal to what we may call the "agnostic" theory of the sidereal world—the theory that investigations into its construction are for ever barred by failure of the means of communication—that we can never see more than a necessarily meaningless part of a possibly infinite, and, in any case, absolutely inscrutable whole.

The general telescopic exploration of the Milky Way began and ended with the Herschels. Their great reflectors have been superseded by the photographic camera. This particular application of its versatile powers encountered special difficulties; but they were happily overcome by Professor Barnard in July, 1889. A six-inch portrait lens afforded the two chief requisites of a powerful light-grasp and an extensive field; and plates exposed with it for some three hours showed accordingly, for the first time, "in all their delicacy and beauty " (to quote Professor Barnard's words), " the vast and wonderful cloud-forms, with their remarkable structure of lanes, holes, and black gaps, and sprays of stars, as no eye or telescope can ever hope to see them." The work has since been continued by him and others, notably by Mr. Russell at Sydney, and by Professor Max Wolf at Heidelberg, so that the complete round of the " circling zone " will, before long, have its varied aspects permanently recorded. They frequently present strange and significant forms. Branching, leaf-like, spiral, elliptical structures abound; individual stars are disposed in circlets, streams, parallel rows, curves of sundry kinds. A " clustering power " of unknown nature is ubiquitously active ; orderly development is in progress. A creative purpose can be *felt*, although it cannot be distinctly followed by the mind.

Herschel's " sweeps " in southern skies were continued until January, 1838; but with frequent intermissions. He was ready for every interesting object that came in his way—comets among the rest. " Encke's—*yours*," he informed his aunt, October 24 1835, " escaped me owing to trees and the Table Mountain, though I cut away a good gap in our principal oak avenue to get at it." Four days later he caught sight of Halley's comet at its second predicted return. But for the stellar aspect of this body his observations of it would have begun much earlier ; for, in the absence of an exact ephemeris, it was impossible to pick it out from among the stars it long precisely counterfeited. " I am sure," he said, " that I must often have swept with a night-glass over the very spot where it stood in the mornings before sunrise; and never was surprise greater than mine at seeing it riding high in the sky, broadly visible to the naked eye, when pointed out to me by a note from Mr. Maclear, who saw it with no less amazement on the 24th."

"This comet," he wrote to Miss Herschel, March 8, 1836, "has been a great interruption to my sweeps, and I *hope* and *fear* it may yet be visible another month." It lingered on just two. He watched with astonishment the changes it underwent. "Within the well-defined head," he wrote in his "Cape Observations," "and somewhat eccentrically placed, was seen a vividly luminous nucleus, or rather, an object which I know no better way to describe than by calling it a miniature comet, having a nucleus, head, and tail of its own, perfectly distinct, and considerably exceeding in intensity of light the nebulous disc or envelope."

This strangely organised body was a very Proteus for instability of form. It alternately lost and recovered its tail. It contracted into the likeness of a star, then dilated into a nebulous globe, which at last vanished as if through indefinite diffusion. The whole mass "seemed touched, seemed turned to finest air." During one week at the end of January—it had passed perihelion November 16—Sir John estimated that the cometary Amœba had increased its bulk no less than forty times !

The paraboloidal form characteristic of this comet and many others, was to him "inconceivable," apart from the play of repulsive, in addition to attractive forces; and he suggested that high electrical excitement due to vaporisation, if of the same kind with a permanent charge on the sun, would plausibly account for the enigmatical appearances he had witnessed. From their close study at Königsberg,

Bessel had already concluded "the emission of the tail to be a purely electrical phenomenon."

In March, 1836, Herschel attacked the subject of southern stellar photometry. Carrying further the "method of sequences," he determined the relative brightness of nearly five hundred stars, which he disposed in order on a single descending scale, and linked on by careful comparisons to the northern stars, as they "lightened into view" on the homeward voyage. By the device of an "artificial standard star," he was besides enabled to obtain numerical values for the lustre of each star examined, in terms of that of Alpha Centauri. Most important of all, he rectified the current system of magnitudes, and introduced a definite "light ratio," which has since been extended, and more strictly defined, but not altered.

His "astrometer" gave Herschel the means of *balancing* the lustre of Alpha Centauri against full moonlight. The latter proved to be 27,500 times more powerful. And Wollaston having determined the ratio of moonlight to sunlight at $\frac{1}{800000}$ (corrected by Zöllner to $\frac{1}{600000}$), it became feasible to compare the brightness of any particular star, *as we see it*, with the brightness of the sun. Alpha Centauri, for example, sends us, according to Herschel, $\frac{1}{22}$ thousand millionth of the light we receive from our domestic luminary. Moreover, when the distance of the star came to be measured (it amounts to twenty-five billions of miles), *light received* could at once be translated into *light emitted*. And the result has been to show that the components of this splendid binary are, taken together, four times more luminous than the sun. Through Sir John Herschel's photometric researches,

then, the real light-power of stars at known distances
became an ascertainable quantity; and it is an element
of great importance to astrophysical inquiries.

On January 10, 1837, he wrote from Feldhausen
to his brother-in-law: " I am now at work on the
spots in the sun, and the general subject of solar
radiation." The sun was just then at an exceptionally
high maximum of disturbance. Spots of enormous
size frequently obscured its disc. One was estimated
by Herschel, March 29, 1837, to cover, independently
of outliers, an area of 3,780 millions of square miles.
So that it considerably exceeded in dimensions the
great spot-group of February, 1892, the largest ever
photographed at Greenwich. The study of a series of
such phenomena led him to propound the " cyclone-
theory " of their origin. It marked a decided advance
in solar physics, if only because it rested upon the
fact—until then unaccountably overlooked—that spot-
production is intimately connected with the sun's
rotation. He regarded it as a kind of disturbance
incidental to a system of fluid circulation analogous
to the terrestrial trade- and anti-trade winds. "The
spots," he said, " in this view of the subject would
come to be assimilated to those regions on the earth's
surface where, for the moment, hurricanes and tornadoes
prevail; the upper stratum being temporarily carried
downwards, displacing by its impetus the two strata
of luminous matter beneath, the upper of course to
a greater extent than the lower, and thus wholly or
partially denuding the opaque surface of the sun
below."

But the fundamental cause of our atmosphere's
flow and counter-flow is absent in the sun. The earth
is heated from the outside, and therefore unequally;

hence the air rushes along, turning westward as it goes, from the chilly poles to the torrid zone of vertical sunshine. No reason is, however, apparent why the solar equator should be hotter than the solar poles. That adduced by Herschel is certainly inadequate. He supposed that, by a retention of heat at the equator due to the accumulation there, consequent upon his rotation, of the sun's absorbing atmosphere, a difference of temperature might be maintained sufficient to keep the solar trade-winds blowing. But the effect is too slight to be detected. And, in fact, the main drift of the photospheric layers is along parallels of latitude. Polar and equatorial currents are insignificant and uncertain.

Herschel and Pouillet contemporaneously, although at opposite sides of the globe, succeeded in 1837 in measuring the intensity of solar radiation. They were the first to apprehend the true bearings of the question, which in principle are simple enough. All that is required is to determine the heating effects, in a given time, of direct sunshine. Its despoilment by our air has, indeed, to be allowed for. Here the chief element of uncertainty comes in. Herschel put the loss at one-third the original thermal power of vertical rays; Pouillet pronounced it nearly one-half; Langley, using the most refined appliances, concludes it to be four-tenths. Striking an average between his own and the French results, Herschel calculated that, at the sun's surface, a shell of ice forty feet thick would melt in one minute, the rate being reduced, at the distance of the earth, to an inch in two hours and twelve minutes. And it is now practically certain that this estimate was too small by about half its amount.

By way of illustrating the effects obtained with his philosophical apparatus, he constructed a popular kind of actinometer, in the shape of an "American dispatch," made of a few pieces of wood and two panes of glass, in which eggs were roasted, and beef-steaks broiled, by sun-heat alone. The viands thus *cosmically cooked* were "eaten with no small relish by the entertained bystanders."

Mimas and Enceladus, Saturn's innermost moons, had persistently eluded Herschel's search for them in England; but, to his great delight, both favoured him at the Cape. His observations of them in 1835-6 were the first since his father's time. The next detection of Mimas was by Mr. Lassell in 1846.

The extent, variety, and completeness of the work done at Feldhausen strike one with ever-fresh admiration. It seems scarcely credible that so much was accomplished in four years by a single unaided individual. Herschel's only assistant was an honest mechanic named John Stone, faithful, serviceable, in his way skilful, but not a "being" of the "quick as lightning" sort, imagined and realised by Caroline Herschel. It is related that during his observations of Halley's comet, Sir John on one occasion fell asleep, and while he remained in this condition of peril (owing to the elevation and insecurity of his perch), Stone kept dutifully turning the telescope. At last the astronomer awoke, rubbed his eyes, looked down the great tube, saw nothing, rubbed his eyes again, and exclaimed, "Why, John, where's the comet?" The comet had meantime set, and the telescope was duly directed towards its place behind Table Mountain!

The splendid fulfilment of his astronomical tasks did not represent the whole of Herschel's activity at the Cape. He collected a large store of tidal data for Dr. Whewell; started scientific meteorology; established a system of national education still working beneficially, and presided over the South African Literary and Scientific Institution, the members of which presented him with a gold medal on his departure. His visit made an epoch in the development of the Colony.

To himself personally it was a time of intense enjoyment. His labours, arduous though they were, proceeded calmly, disembarrassed from jostling claims and counter-claims. They were carried on with absorbed enthusiasm, inspired in part by their sublime nature, in part by the excitement of novelty. His family throve and multiplied at Feldhausen. Sir Thomas Maclear's friendship supplied unfailing social pleasure. An exhilarating climate, moreover, enchanting scenery, translucent skies, blossoming glens and hillsides worthy of Maeldune's Isle of Flowers, contributed to render his southern sojourn a radiant episode. He wrote of it to Mr. Stewart as " the sunny spot in my whole life, where my memory will always love to bask." But " the dream," he added, " was too sweet not to be dashed by the dread of awakening." The spell was broken when in the middle of March, 1838, he sailed in the *Windsor Castle* for England.

The interest created by his romantic expedition spread to the other side of the Atlantic. A grotesque narrative, published in the *New York Sun* for September, 1835, of lunar discoveries made at the Cape with the combined aid of the twenty-foot reflector and the Drummond limelight, was eagerly read and believed by thousands, was reprinted, re-circulated, and re-read. Nor were common gulls the only victims to the hoax. The truth of the story was gravely debated by the Paris Academy of Sciences.

Herschel's home-coming was a triumph. He was overwhelmed with applause and gratulation. His fellow-countrymen offered him what compensation they could for the disappearance from his horizon of the Southern Cross. He was created a baronet at the Queen's Coronation, received an honorary degree of D.C.L. at Oxford in 1839, and was offered, but declined, reimbursement from the Treasury for the entire cost of his trip. He peremptorily refused as well to represent the University of Cambridge in Parliament, or to be nominated for the Presidentship of the Royal Society. His utmost desire was for a quiet and laborious life. A banquet, however, given in honour of his return, June 15, 1838, could not be shunned; the less so that the celebration had a typical character. "In honouring a man," Sir William Hamilton said, in proposing his health, "we honour science too." For "the cultivators and lovers of Science have chosen Herschel for their chief—say, rather, have as such received him by inheritance."

CHAPTER IX.

HERSCHEL'S career as an observing astronomer came
to a virtual end with his departure from the Cape.
He was then forty-six, two years younger than his
father when he began his course of prodigious activity
at Slough. Sir William's craving to see and to know
was insatiable ; Sir John's was appeased by the
accomplishment of one grand enterprise. His was a
many-sided mind ; dormant interests of sundry kinds
revived on the first opportunity ; new ones sprang up ;
and curiosity to interrogate the skies ceased to " prick
the sides of his intent." So the instruments taken
down at Feldhausen in 1838 were not remounted in
England ; and their owner is never again recorded to
have used a telescope. One cannot but regret that, in
the plenitude of his powers, and instructed by rare
experience, he should have put by his weapons of
discovery.* The immense stock of observations with
which they had furnished him remained, it is true, in
their primitive, rough-hewn state ; and he may have
considered that wise husbandry required him to save
one harvest before planting another. This, at any
rate, was the course that he pursued.

But it was often and in many ways interrupted.
The demands on his time and thoughts were innumer-

* The three specula of the twenty-foot are in the possession of
Sir William J. Herschel ; the tube remains in good preservation at
Collingwood.

able. Having settled his family for the season in
London, he paid his third and last visit to his venerable aunt, and, in returning, dined with Dr. Olbers,
the physician-astronomer of Bremen, then in his
eightieth year. A fortnight later he was on his way
to Newcastle, where the British Association met,
August 20th. He was received with acclamation, but
overwhelmed by scientific exactions. The proceedings
were to him " a dreadful wear and tear," and they
left behind " mixed and crowded recollections." No
wonder. Besides acting as President of the Mathematical Section, he found himself involved in varied
responsibilities. He was placed on a Committee for
bringing down to date the places of Lacaille's 10,000
southern stars ; on another for revising stellar nomen-
clature. The reduction of a body of meteorological
observations made on a plan of his devising was
entrusted to him ; above all, he was charged with the
development of Humboldt's international scheme for
securing systematic and world-wide observations on
terrestrial magnetism. He drew up a memorial to
the Government ; compiled the Instructions for Sir
James Clark Ross's Antarctic expedition ; and elabor-
ately reported progress at several successive meetings
of the British Association. His heart was in the work.
He contributed an article dwelling on its importance
to the *Quarterly Review* for June, 1840; and in
1845 he expressed the opinion that " terrestrial physics
form a subject every way worthy to be associated with
astronomy as a matter of universal interest and public
support."

The constellations gave him still more trouble
than the vagaries of poised needles. They were in a
riot of disorder. Celestial maps had become " a system

of derangement and confusion"—of confusion "worse confounded." New asterisms carved out of old existed precariously, recognised by some, ignored by others; waste places in the sky had been annexed by encroaching astronomers as standing-ground for their glorified telescopes, quadrants, sextants, clocks; a chemical apparatus had been set up by the shore of the river Eridanus, itself a meandering and uncomfortable figure; while serpents and dragons trailed their perplexing convolutions through hour after hour of right ascension. There were constellations so large that Greek, Roman, and Italic alphabets had been used up in designating the included stars; there were others separated by debatable districts, the stars in which often duplicated those situated within the authentic form of one of the neighbouring celestial monsters. Identification was thus in numberless cases difficult; in some, impossible.

In conjunction with Francis Baily, Herschel undertook the almost hopeless task of rectifying this intolerable disorder. After much preliminary labour, he submitted to the Royal Astronomical Society, in 1841, a drastic scheme of constellational reform—a stellar redistribution-bill, framed on radical principles. Its alarming completeness, however, caused it to be let drop; and he finally proposed, in his report of 1844 to the British Association, a less ambitious but more practicable measure. Although not adopted in its entirety, it paved the way for ameliorations. The boundaries of the constellations have since been defined; interlopers have been ejected; one—the Ship Argo —especially obnoxious for its unwieldy dimensions, has been advantageously trisected. Nevertheless, individual star-nomenclature grows continually more

M

perplexed; partial systems have become intermingled and entangled; double stars are designated in one way, variables in another, quick-moving stars in a third, red stars in a fourth, while any one of many catalogue-numbers may be substituted at choice; palpable blunders, unsettled discrepancies, anomalies of all imaginable kinds, survive in an inextricable web of arbitrary appellations, until it has come to pass that a star has often as many aliases as an accomplished swindler.

In the spring of 1840 Herschel removed from Slough to Collingwood, a spacious country residence situated near Hawkhurst, in Kent.   Here he devoted himself, in good earnest, to the preparation of his Cape results for the press.   It was no light task.   The transformation of simple registers of sweeps into a methodical catalogue is a long and irksome process; and Herschel was in possession of the "sweepings" of nearly four hundred nights.   He executed it single-handed, being averse to the employment of paid computers.   This was unfortunate. Monotonous drudgery was not at all in his line; as well put Pegasus between shafts.   He had always found in himself "a great inaptitude" for numerical calculations; and he now acknowledged to Baily that attention to figures during two or three consecutive hours distressed him painfully.   Whewell lamented in the *Quarterly Review* the lavish expenditure of his time and energy upon "mere arithmetic"—computations which a machine would have been more competent to perform than a finely organised human brain. At last, however, in November, 1842, the necessary reductions were finished; and the letterpress to accompany the catalogues of double stars and nebulæ left his hands a couple of years later. The preparation of the plates occasioned further vexatious delays ; and it was not until 1847 that the monumental work entitled "Results of Astronomical Observations at the Cape of Good Hope" issued from the press. The expenses of its production were generously defrayed by the Duke of Northumberland. In sending a copy to his aunt, then in her ninety-eighth year, he wrote : "You will have in your hands the completion of my father's work—'The Survey of the Nebulous Heavens.'" The publication was honoured with the Copley Medal by the Royal Society, and with a special testimonial by the Astronomical Society.

Bessel, the eminent director of the Königsberg observatory, made Herschel's personal acquaintance at the Manchester meeting of the British Association in 1842, and paid him a visit at Collingwood. The subject of a possible trans-Uranian planet was discussed between them. The German astronomer regarded its existence as certain, and disclosed the plot he had already formed for waylaying it on its remote path. The premonition stirred Herschel deeply. "There ought to be a hue and cry raised!" he exclaimed in a letter to Baily. And in resigning the Chair of the British Association, September 10, 1846, he spoke with full assurance of the still undiscovered body. "We see it," he declared, "as Columbus saw America from the shores of Spain. Its movements have been felt, trembling along the far-reaching line of our analysis, with a certainty hardly inferior to that of ocular demonstration." Within a fortnight, Neptune, through Le Verrier's indications, was captured at Berlin.

M 2

" I hope you agreed with me," he wrote, November 19, 1846, to Sir William Hamilton, " that it is perfectly possible to do justice to Adams's investigations without calling in question M. Le Verrier's *property* in his discovery. The fact is, I apprehend, that the Frenchmen are only just beginning to be aware *what a narrow escape Mr. Neptune had of being born an Englishman.* Poor Adams aimed at his bird, it appears, first, and as well as Le Verrier, but his gun hung fire, and the bird dropped on the other side of the fence ! "

It is well known that Le Verrier and Adams personally ignored controversy as to their respective claims to the planetary *spolia opima.* They were together at Collingwood in July, 1847, with Struve as their fellow-guest. During those few days King Arthur (in the person of Sir John Herschel) " sat in hall at old Caerleon."

He was elected President of the Royal Astronomical Society for the usual biennial term in 1828, 1840, and 1847; on the last occasion through the diplomatic action of Professor De Morgan. The Society was passing through a crisis ; he apprehended its dissolution, and judged that it could only be saved by getting Herschel's consent to become its nominal head. " The President," he wrote to Captain Smyth, " must be a man of brass (practical astronomer)—a micrometer-monger, a telescope-twiddler, a star-stringer, a planet-poker, and a nebula-nabber. If we give bail that we won't let him do anything if he would, we shall be able to have him, I hope. We must all give what is most wanted, and his name is even more wanted than his services. We can do without his services, not without loss, but without difficulty. I see we shall not, without great difficulty, dispense with his name."

And to Herschel himself: "We have been making our arrangements for the Society for the ensuing year; and one thing is that you are not to be asked to do anything, or wished to do anything, or wanted to do anything. But we want your *name*." It was lent; and its credit seems to have had the desired effect.

Dr. Whewell vainly tried to inveigle him, in November, 1838, into accepting the presidentship of the Geological Society; but he had to submit, in 1842, to be elected Rector of Marischal College, Aberdeen; and he consented to preside over the meeting of the British Association at Cambridge in June, 1845. His dignity on the occasion was not allowed to interfere with his usefulness. He wrote home June 22: "We have been on the Magnetic Committee working hard all the morning, in a Babel of languages and a Babylonian confusion of ideas, which crystallised into something like distinctness at last." By that time the long-desired particulars regarding terrestrial magnetism were rapidly accumulating. *Facts,* as Herschel announced from the Presidential Chair, were plentifully at hand. "What we now want is *thought,* steadily directed to single objects, with a determination to avoid the besetting evil of our age—the temptation to squander and dilute it upon a thousand different lines of inquiry."

Herschel observed the great comet of 1843 from the roof of his house at Collingwood, on March 17, the first evening of its visibility in England. All that could be seen was "a perfectly straight narrow band of considerably bright, white cloud, thirty degrees in length, and about one and a half in breadth." It was not until the following night that he recognised in this strange "luminous appearance" "the tail of a magnificent comet, whose head at the times of both observations was below the horizon."

In December, 1850, he was appointed Master of the Mint—a position rendered especially appropriate to him by Newton's prior occupation of it. The duties connected with it were just then peculiarly onerous. Previously of a temporary and political character, the office now became permanent, and simply administrative. Many other changes accompanied this fundamental one. "The whole concern," he said, "is in process of reorganisation." This fresh start demanded much "personal and anxious attendance." Notwithstanding his anxious regard for the interests of subordinates, the reconstruction could not but be attended by serious friction. No amount of oiling will get rusty wheels to revolve smoothly all at once. "Things progress rather *grumpily*," he reported privately, "owing to the extreme discontent of some parties." Further contentious business devolved upon him as a member of the jury on scientific instruments at the Great Exhibition. His time was fully and not agreeably occupied. Rising at six, he worked at home until half-past nine, then hurried to the Mint, which he exchanged between three and four o'clock for the Exhibition, and there, until the closing of its doors, examined the claims, and appeased the quarrels of rival candidates for distinction. He also sat on the Royal Commission appointed in 1850 to inquire into the University system. Its recommendations, agreed to by him in 1855, greatly disgusted Whewell; but their friendship remained unaltered by this discordance of opinion.

These accumulated responsibilities were too much for Herschel's sensitive nature; and the burthen was made heavier by a partial separation from his family. He was never alone in Harley Street, but the joyous life of Collingwood could not be transported thither; and the arid aspect of a vast metropolis, suggesting business and pleasure in excess, but little of enjoyment in either, oppressed him continually. His health suffered, and in 1855 he withdrew definitively into private life. His resignation of the Mint was most reluctantly accepted.

"I find," playfully remarked De Morgan, "that Newton and Herschel added each one coin to the list: Newton, the gold quarter-guinea, which was in circulation until towards the end of the century; Herschel, the gold quarter-sovereign, which was never circulated."

It was not the repose of inaction that Herschel sought at Collingwood. "Every day of his long and happy life," Professor Tait said truly, "added its share to his scientific services." Thenceforward he devoted himself chiefly to the formidable task of collecting and revising his father's results and his own. His "General Catalogue of Nebulæ," published in the *Philosophical Transactions* for 1864, was in itself a vast undertaking. It comprised 5,079 nebulæ and clusters, to which it served as a universal index of reference. It averted the mischief of duplicate discoveries, settled the sidereal status of many a pseudo-comet, and quickly became the authoritative guide

of both comet- and nebula-hunters. In the enlarged
form given to it by Dr. Dreyer in 1888, it is likely
long to hold its place. Herschel next, in 1867, amal-
gamated into a regular catalogue of 812 entries his
father's various classed lists of double stars (Memoirs,
Royal Astronomical Society, xxxv.). A far more com-
prehensive work was then taken in hand. He desired
to do for double stars what he had done for nebulæ
—to compile an exhaustive register of them in the
shape of a catalogue, accompanied by a short descriptive account of each pair. But he was not destined
to put this coping-stone to the noble monument
erected by his genius. Strength failed him to digest
and dispose the immense mass of materials he had
collected. Nor was it possible for another to gather
up the loose threads of his unfinished scheme. All
that could be done was to preserve the imposing
fragment as he left it. An ordered list of the 10,320
multiple stars he had proposed to treat was accordingly published in the fortieth volume of the same
Society's *Memoirs* under the care of Professor
Pritchard and Mr. Main. But it hardly possesses
more than a commemorative value.

Maria Edgeworth was an old friend of Sir John
Herschel's. In March, 1831, she paid him a three
days' visit at Slough, which, she told a friend in Ireland,
"has far surpassed my expectations, raised as they
were, and warm from the fresh enthusiasm kindled by
his last work" (the "Preliminary Discourse"). Mrs.
Herschel she described as "very pretty," sensible, and
sympathetic, and possessed of the art of making
guests happy without effort. On Sunday, after service, the philosopher showed off the dazzling colour-
effects of polarised light, and at night, with the twenty-

foot, " Saturn and his rings, and the moon and her volcanoes."

After twelve years, she came again, this time to Collingwood. " I should have written before," Herschel assured Sir William Hamilton, December 1st, 1843, " but Miss Edgeworth has been here, and that, among all people who know how to enjoy her, is always considered an excellent reason for letting correspondence and all other worldly things ' gang their ain gate.' She is more truly admirable now, I think, than at any former time, though in her seventy-fifth year."

Maria herself wrote from Collingwood in the following spring : " Here are Lord and Lady Adare, Sir Edward Ryan, and ' Jones on Rent.' Jones and Herschel are very fond of one another, always differing, but always agreeing to differ, like Malthus and Ricardo."

Sir Willliam Hamilton spent a week under the same hospitable roof in 1846. He was delighted, and, as was his wont, compressed the expression of his pleasure " within the Sonnet's scanty plot of ground." In the first of a pair entitled " Recollections of Collingwood," he celebrated the " thoughtful walk " with his host, and the " social hours " in a family circle,

> " Where all things graceful in succession come ;
> Bright blossoms growing on a lofty stalk,
> Music and fairy-lore in Herschel's home." *

The second dealt with " high Mathesis," and

> " dimly traced Pythagorean lore ;
> A westward-floating, mystic dream of FOUR."

* The lines are quoted in Graves's " Life of Hamilton," vol. ii. p. 525.

Although not, like his friend, an incorrigible and impenitent sonnetteer, Herschel was "very guilty" of at least one specimen of the art. They were staying together, in June, 1845, at Ely, in the house of Dean Peacock. Hamilton's inevitable sonnet came duly forth, and "next morning," he related to De Morgan, "as my bedroom adjoined Herschel's, and thin partitions did my madness from his great wit divide, I easily heard what Burns might have called a 'crooning,' and was not much surprised (being familiar with the symptoms of the attack)* when, before we sat down to breakfast at the Deanery, Lady Herschel handed me, in her husband's name and her own, a sonnet of *his* to *me*, which, unless the spirit of egotism shall seize me with unexpected strength, I have no notion of letting you see."

The circulation of Herschel's fervid eulogy would assuredly have put his modesty to the blush. Headed "On a Scene in Ely Cathedral," it runs as follows:—

"The organ's swell was hushed, but soft and low
An echo, more than music, rang; when he,
The doubly-gifted, poured forth whisperingly,
High-wrought and rich, his heart's exuberant flow
Beneath that vast and vaulted canopy.
Plunging anon into the fathomless sea
Of thought, he dived where rarer treasures grow,
Gems of an unsunned warmth and deeper glow.
O born for either sphere! whose soul can thrill
With all that Poesy has soft or bright,
Or wield the sceptre of the sage at will
(That mighty mace which bursts its way to light),
Soar as thou wilt!—or plunge—thy ardent mind
Darts on—but cannot leave our love behind."

Of Hamilton's abstruse invention, the method of "Quaternions" (here alluded to), Herschel was, from

* "Aut insanit, aut versos facit."

the first, an enthusiastic admirer. He characterised it in 1847 as "a perfect cornucopia, from which, turn it on which side you will, something rich and valuable is sure to drop out." The "power and pregnancy" of the new calculus were supremely delightful to him, and he advised every mathematician to gain mastery over it as a "working tool." As such it has not yet been brought into ordinary use, yet it remains in the armoury of science, ready for emergencies.

Miss Mitchell of Nantucket, the discoverer of a comet, and a professor of astronomy, published in 1889 (in the *Century* magazine) her reminiscences of a short stay at Collingwood in 1858. Her host " was at that time sixty-six, but he looked much older, being lame and much bent in his figure. His mind, nevertheless, was full of vigour. He was engaged in rewriting the 'Outlines of Astronomy.'" "Sir John's forehead," ·she says, "was bold but retreating; his mouth was very good. He was quick in motion and in speech. He was remarkably a gentleman; more like a woman in the instinctive perception of the wants and wishes of a guest."

"In the evening," she relates, "we played with letters, putting out charades and riddles, and telling anecdotes, Sir John joining the family party and chatting away like the young people." He propounded the question : If one human pair, living in the time of Cheops, had doubled, and their descendants likewise, once every thirty years, could the resulting population find room on the earth ? The company thought not. "But if they stood closely, and others stood on their shoulders, man, woman, and child, how many layers would there be ?" "Perhaps three," replied Miss

Mitchell. "How many feet of men?" he insisted.
"Possibly thirty." "Enough to reach to the moon,"
said his daughter. "To the sun," exclaimed another.
"More, more!" cried Sir John, exulting in the general
astonishment. "To Neptune," was the next bid.
"Now you burn," he allowed. "*Take one hundred
times the distance of Neptune, and it is very near.*"
"That," he added, "is my way of whitewashing war,
pestilence, and famine."

He further entertained his American guest with
accounts of the paradoxical notions communicated to
him by self-taught or would-be astronomers. One
had inferred the non-existence of the moon from
Herschel's chapters on lunar physics and motions.
Another enclosed half-a-crown for a horoscope. A
third wrote, "Shall I marry, and have I seen her?"
In reference to the efforts then being made to intro-
duce decimal coinage into England, he remarked,
"We stick to old ways, but we are not cemented to
them."

The portrait of Caroline Herschel, painted by
Tielemann in 1829, which she herself declared to
"look like life itself," hung in the drawing-room. (It
is that reproduced in this volume.) "You would say
in looking at it," Miss Mitchell wrote, 'she must have
been handsome when she was young.' Her ruffled
cap shades a mild face, whose blue eyes were even
then full of animation. But it was merely the beauty
of age."

Herschel was no exception to the rule that
astronomers love music and flowers. He was never
tired of gardening, and—to quote James Nasmyth—
"his mechanical and manipulative faculty enabled
him to take a keen interest in all the technical arts

which so materially aid in the progress of science."
The manufacture of specula naturally came home to
him, and he watched with genuine pleasure Nasmyth's
grinding and polishing operations. He spent several
days with him at Hammerfield in 1864. "Of all
the scientific men I have had the happiness of
meeting," Nasmyth wrote in his "Autobiography,"
"Sir John stands supremely at the head of the
list. He combined profound knowledge with perfect humility. He was simple, earnest, and com-
panionable. He was entirely free from assumptions
of superiority, and, still learning, would listen atten-
tively to the humblest student. He was ready
to counsel and instruct, as well as to receive
information."

Herschel's correspondence with De Morgan extended over nearly forty years, and became latterly of
an intimate character. "Looking back on our long
friendship," he wrote to the widow shortly after De
Morgan's death in the spring of 1871, "I do not find a
single point on which we failed to sympathise; and I
recall many occasions on which his sound judgment
and excellent feeling have sustained and encouraged
me. Many and very distinct indications tell me that
I shall not be long after him."

It fell out as he had predicted. The obituary
memoirs of the two are printed close together in the
Astronomical Society's "Monthly Notices." After a
prolonged decline of strength, Sir John Herschel
died at Collingwood, in his seventy-ninth year, May
5th, 1871, his intellect remaining unclouded to
the last. He was buried in Westminster Abbey,
near the grave of Newton. The words engraven
above his resting-place, "Coelis exploratis, hic prope

Newtonum requiescit," tell what he did, and what he deserved.

His death created an universal sense of sorrow and of loss. He left vacant a place which could never be filled. His powers, his qualities, and his opportunities made a combination impossible to be reproduced. His genius showed curious diversities from his father's. He lacked his profound absorption, his penetrating insight, his unaccountable intuitions. A tendency to discursiveness, happily kept in check by strength of will and devotion to an elevated purpose, replaced in him his father's enraptured concentration. On the other hand, his appreciative instinct for the recondite beauty of mathematical conceptions was wanting to his father. William Herschel possessed fine mathematical abilities; but he cultivated them no further than was necessary for the execution of his designs; and elementary geometry served his turn. But Sir John might have taken primary rank as a pure mathematician. Possibly his inventive faculty would have developed in that line more strongly than in any other. The grasp of his mind was indeed so wide that many possibilities of greatness were open to him. That he chose rightly the one to make effectivè, no one can doubt. The neglect on his part of astronomy would have been a scientific delinquency. His splendid patrimony of telescopic results and facilities was inalienable. It was a talent entrusted to him, which he had not the right to bury in the ground. He laboured with it instead to the last farthing. Not for his own glory. He aspired only to fill up, for the honour of his father's name, the large measure of his achievements. In doing so he performed an unparalleled feat. He swept from pole to pole the entire surface of the hollow sphere of the sky. It is unlikely to be repeated. The days of celestial pioneering are past. Nothing on the scale of a general survey will in future be undertaken except with photographic help. The use of the direct telescopic method tends to become more and more restricted. This is a loss as well as a gain. A *hortus siccus* is to a blooming garden very much what a collection of photographs is to the luminous flowers of the sky. They are depicted more completely, more significantly, more conveniently for purposes of investigation, than they can be seen ; but the splendour of them is gone. Their direct contemplation has an elevating effect upon the mind, which indirect study, however diligent and instructive, is incapable of producing. The sublimity of the visions drawn from the abyss of space cannot be reasoned about. It strikes home to the spectator's inner consciousness without waiting for the approval of his understanding. Thus to Herschel, no less expressly than to the Psalmist three thousand years earlier, "the heavens told the glory of God." He lived at his telescope a life apart, full of incommunicable experiences.

"To Herschel," as Mr. Proctor expressed it, "astronomy was not a matter of right ascension and declination; of poising, clamping, and reading-off; of cataloguing and correcting." "It was his peculiar privilege," Dean Stanley remarked in his funeral sermon, "to combine with those more special studies such a width of view and such a power of expression as to make him an interpreter, a poet of science, even beyond his immediate sphere." Hence the popularity of his books, and the favoured place he occupied in public esteem.

His character was of a more delicate fibre than his father's. It was also, by necessary consequence, less robust. Sir William Herschel surmounted adversity. Sir John would have endured it, had his lot been so appointed. But it never came his way. He was one of those rarest of rare individuals—

> "Whose even thread the Fates spin round and full,
> Out of their choicest and their whitest wool."

His life was a tissue of felicities. For him there was no weary waiting, no heart-sickening disappointment, no vicissitudes of fortune, no mental or moral tempests. Success attended each one of his efforts ; he could look back without regret ; he could look forward with confident hope ; his family relations brought him unalloyed happiness. He suffered, indeed, one bereavement in the untimely death of his daughter, Mrs. Marshall, the wife of a nephew of Dr. Whewell ; but Christian resignation sweetened his sorrow. His religion was unpretending and efficacious. No duty was left by him unfulfilled ; and he wore, from youth to age, "the white flower of a blameless life." A discriminating onlooker said of him, that his existence "was full of the serenity of the sage and the docile innocence of a child." He was retiring almost to a fault, careless of applause, candid in accepting criticism. Although habitually indulgent, he was no flatterer. "Anyone," Mr. Proctor said, "who objected to be set right when in error, might well be disposed to regard Sir John Herschel as a merciless correspondent, notwithstanding the calm courtesy of his remarks. He set truth in the first place, and by comparison with her, neither his own opinions, nor those of others, were permitted to have any weight whatever." Beginners invariably met with his sympathy and encouragement. He felt for difficulties which he himself had never experienced.

Being thus constituted, he could not but inspire affection. The French physicist, Biot, when asked by Dr. Pritchard, after the death of Laplace, who, in his opinion, was his worthiest successor, replied, "If I did not love him so much, I should unhesitatingly say, John Herschel." His own attachments were warm and constant; and the few scientific controversies in which he engaged, were carried on with his habitual gentleness and urbanity.

Herschel left eight daughters and three sons, of whom the eldest, Sir William James Herschel, succeeded him in the baronetcy, while the second, Professor Alexander Stewart Herschel, has earned celebrity by his meteoric researches. The election of the third, Colonel John Herschel, to a Fellowship of the Royal Society, in recognition of his spectroscopic examination of southern nebulæ, threw a gleam of joy over his father's deathbed. Lady Herschel survived her husband upwards of thirteen years.

The learned societies of Europe vied with each other in enrolling the name of Sir John Herschel; and he was nominated, in 1855, on the death of Gauss, one of the eight foreign members of the French Academy of Science. As we have seen, he received the Copley Medal from the Royal Society twice, their Royal Medal thrice, and from the Royal Astronomical Society, two Gold Medals and a testimonial. Compliments and homage, however, left him as they found him—quiet, intent, and unobtrusive.

Several portraits of him are in existence. One was executed in oils by Pickersgill for St. John's College,

N

Cambridge, at a comparatively early period of his life. It is here (page 142) reproduced from an admirable engraving. His later aspect is finely represented in a painting by his eldest daughter, Lady Gordon. The eyes in it are sunken, though brilliant; the shape of the head is concealed by a mane of grey hair. There is about it something of leonine grandeur, disjoined from leonine fierceness. It perpetuates, indeed, the countenance of a man replete with human tenderness.

CHAPTER X.

WRITINGS AND EXPERIMENTAL INVESTIGATIONS.

COULD the whole of Sir John Herschel's astronomical career be obliterated, and the whole of his contributions to pure mathematics be forgotten, he would still merit celebrity as a physicist. Experimental optics, above all, engaged his attention. " Light," he himself said, " was his first love," and he was never wholly forgetful of it. In 1830 he described himself as " forcibly drawn aside from his optical studies" by the claims of nebulæ and double stars. How strong he felt those claims to be, can best be understood by considering the firmness with which he averted his mind, out of regard to them, from the intricate and bewitching subject of his early devotion.

" I understand from Peacock," Dr. Whewell wrote to him, June 19, 1818, " that you are untwisting light like whipcord, examining every ray that passes within half a mile, and putting the awful question, 'Polarised, or not polarised ? ' to thousands that were never before suspected of any intention but that of moving in a straight line." These interrogatories brought out a remarkable diversity in the action upon light of quartz, and other similar substances, corresponding with the two different modes of crystallisation belonging to each of them. Here, in Lord Kelvin's phrase, is " one of the most notable meeting-places between natural history and natural philosophy."

N 2

The nascent science of spectrum analysis was materially promoted by Herschel. He noticed in 1819 the distinctive light-absorbing qualities of coloured media, studied the spectra of various flames, adverted to the definiteness and individuality of the bright lines composing them, and recommended their employment for purposes of chemical identification.

A year later, he developed and modified Brewster's explanation of the colours of mother-of-pearl. They do not, like the iridescence of a fly's wing, result from the interference of waves of light reflected from two closely adjacent surfaces, but from interference brought about by the finely striated texture of the shell's surface, and a cast of the rainbow-tinted surface in black sealing-wax will display the same sheen of colour as the original. Herschel detected, however, a second more closely striated structure which cannot be impressed upon plastic matter.

Up to this time he accepted unreservedly the emission theory of light. But a candid study of Young's and Fresnel's writings produced a fundamental change in his opinions; and in an article on " Light," written for the " Encyclopædia Metropolitana " in 1827, he expounded the undulatory theory with all the ardour of a neophyte. He brought thereby one of the grandest generalisations of science into universal currency, and enforced its acceptance by the cogency of his arguments, the logical order of his method, and the lucidity of his style. The treatise was translated into French by Quetelet; and no reader, Professor Pritchard remarked, " could escape the charm of the half-suppressed enthusiasm which carried him along."

Whewell ranked him "among the *very* small number of those who, in the singularly splendid and striking researches of physical optics, had both added important experimental laws to those previously known, and weighed the relations of these discoveries to the refined and recondite theory towards which they seemed to point." He contributed to the same Encyclopædia scarcely less brilliant essays on Heat, Sound, and Physical Astronomy.

"Do not observe too much in cold weather," Miss Herschel advised her nephew, in anticipation of the winter of 1831-2; "write rather books to make folks stare at your profound knowledge."

He followed the positive part of her counsel. Indeed, his "Preliminary Discourse on the Study of Natural Philosophy" had made its appearance in the previous year, as the introductory volume to Lardner's "Cabinet Cyclopædia." It was greeted with a chorus of approbation. Gauss reviewed it in the *Gelehrte Anzeigen*, Whewell in the *Quarterly Review*. Translated into French, German, and Italian, it delighted "all sorts and conditions" of readers with the justice and breadth of the views set forth in it agreeably, easily, and without pretension to superiority. The book included a survey of the actual state of scientific knowledge, and a philosophy of its augmentation. Students derived from it, Gauss remarked, both information as to how accepted results had been obtained, and guidance for their personal investigations. Herschel was exceptionally qualified, Whewell wrote, "to expound the rules and doctrines of that method of research to which modern science has owed its long-continued, steady advance, and present flourishing condition." He had the knowledge, without the narrowness, of a specialist in almost every department of experimental physics.

"With singular alacrity," he came to the front wherever there seemed a chance of pushing back the barriers of ignorance. A disciple of Bacon, he had the advantage over his master of being habitually conversant with the practical working of inductive methods. The treatise was styled by Whewell "an admirable comment on the 'Novum Organum.'" One, however, possesses the indefinable quality of *greatness;* it stands out from the centuries a solid structure, clothed with visionary magnificence; the other is elegant, attractive, wise, acute, even profound, but not in any degree, or from any point of view, *great.*

It was followed, in 1833, by "A Treatise on Astronomy," published in the same series. An "Edinburgh Reviewer" (doubtless Brougham once more) perused it with regret. "The proper position of Sir John Herschel" he considered to be "at the head of those who are nobly, though it may be silently and without notice, endeavouring to extend the present limits of human knowledge," rather than among "the ranks of those whose office it is to herald the triumphs of science, and point out its treasures and results to the admiration of the vulgar." This ostensibly flattering estimate was made the basis for an imputation of vanity. The inducements, according to the critic, were strong "to descend from the airy summits of abstract science to the level at which the great body of the reading public can appreciate and applaud. Philosophers, like other writers, naturally wish to be read, and to have reputation; and reputation, as was remarked by d'Alembert, depends more upon the number than the merit of those who praise." Sir John Herschel would have been better employed in pursuing the track of original discoveries, leaving new truths to " find their way to the drawing-room as best they might." The whole tenour of his life refuted these insinuations.

The " Treatise on Astronomy " was enlarged in 1849 into the deservedly famous " Outlines of Astronomy." Twelve editions of this book were published, the last in 1873 ; it was translated into Chinese and Arabic, as well as into most European languages, including Russian ; it made a profound and lasting impression upon the public mind. No science has perhaps ever received so masterly a general interpretation. Methodical in plan, inspiriting in execution, it demands readers willing to share some part of the pains, for the sake of partaking in the high pleasures of the writer. For it is popular in the sense of eschewing mathematical formulæ, not in the sense of evading difficulties.

The work fittest to be set by its side is the " Exposition du Système du Monde." But Laplace restricted his view to the sun's domain, while Herschel excluded from his no part of the sidereal universe. Laplace was, besides, a geometer in the first, an astronomer only in the second place. The movements of the heavenly bodies interested him because they afforded opportunities for analytical triumphs. Their intricacy notwithstanding, he was elated to find that they could not baffle his ingenuity in constructing formulæ to correspond. Their balance, their harmony, their obedience to a single and simple law, gratified the orderly instincts of his powerful yet frigid mind. Where he could not explain, however, he did not admire. Mystery had no attraction for him. Knowledge, to be knowledge in his eyes, should have definite, clear-cut outlines. His scheme of the

universe was like the map of the world laid down by
Hecatæus, neatly finished off with a circumfluent
ocean-stream; it included no intimations of a *beyond*.
Herschel's, on the contrary, might be compared to the
map of Herodotus, in which some details were filled
in, while the external boundary had been abolished.
The most essential part of the progress made in the
interval consisted in leaving verge and scope for the
unknown.   Next to nothing remained to be learned of
the heavens, as they presented themselves to the
author of the " Mécanique Céleste"; while Herschel
saw everywhere only beginnings, possibilities of discovery, and dim prospects of "ultimate attainments,"
as to the realisation of which "it would be unwise to be
sanguine, and unphilosophical to despair" (Playfair).
At the head of very many of his chapters he might,
without presumption, have written: "Quorum pars
magna fui."   They gave largely the results of his
personal investigations, and were vivified by immediate
acquaintanceship with the objects described.   Hence
the unsought picturesqueness of his descriptive
epithets, and the sublimity of trains of thought communicated to him direct from the unveiled heavens.

Herschel invented in 1825, jointly with Babbage,
the "astatic," or neutralised magnetic needle—a
little instrument which was no sooner available
than it was found to be indispensable. "Nihil tetigit
quod non ornavit."   And many and various were the
things touched by his versatile genius. He had a
narrow escape of becoming for life a chemist. At the
very outset of his career he applied for the vacant chair
of that science at Cambridge; but was left, as he himself humorously expressed it, "in a glorious minority
of one."   The chemical inquiries, nevertheless, which

he carried on at Slough brought to his notice one set of relations of no trifling importance. This was the solvent effect upon salts of silver of the hyposulphites of soda, potash, etc. The discovery was turned to account by himself in 1840 for the "fixing" of photographic images. It secured the future of the embryo art. By the agency of hyposulphite of soda in washing away the unaffected chloride of silver, while leaving untouched the parts of the deposit decomposed and darkened by exposure, permanent light-pictures, capable of indefinite multiplication, were at length secured.

On March 14th, 1839, unaware that he had been anticipated by Fox Talbot, Herschel presented to the Royal Society twenty-three prints made by the sensitised paper process. A memoir communicated in 1840 was full of suggestive novelties. In it he described experiments on "the chemical analysis of the solar spectrum," pointing out that the character and amount of the action exercised by the various rays depend mainly upon the nature of the substance acted upon. He made a start, too, with spectral photography, and his detection of the "lavender-grey" effect to the eye of the ultra-violet section might be said to have added a new note to the prismatic gamut. In the opposite, or infra-red end, by simply letting the solar spectrum fall upon a strip of paper moistened with alcohol, he detected, through the different rates of drying where they fell, some of the "cold bands," by which the invisible heat-rays are furrowed. The photo-spectroscopic apparatus devised for the purpose of these researches formed part of the Loan Collection of Scientific Instruments exhibited at South Kensington in 1876.

Still more essential was the improvement of sub-stituting for paper, glass plates spread with a sensitive film. A photograph of the old forty-foot telescope, taken by this method in 1839, and preserved in the South Kensington Museum, is of unrivalled antiquarian value as regards the history of photography. The terms "positive" and "negative" received in this remarkable paper their now familiar photographic meaning. Its merits were acknowledged in 1840 by the award of a Royal Medal.

Sir John Herschel would, doubtless, at that time have set aside as a chimera the notion that the art he was engaged in promoting was destined, in large measure, to supersede visual methods in astronomy; that the great telescopes of the future would find their most useful employment in concentrating the rays of celestial objects upon sensitive plates. He soon perceived, however, the importance of photo-graphy as an adjunct to direct observation, and recommended, in 1847, the automatic self-registration of sun-spots. This hint—emphasised in 1848—was acted upon in 1858, when the regular collection of documentary evidence as to the sun's condition was begun at Kew with De la Rue's "photoheliograph."

In 1845 he published the first effective investi-gation of "fluorescence," called by him "epipolic," or superficial, "dispersion." This curious phenomenon consists in the illumination to the eye of certain sub-stances, such as sulphate of quinine and canary glass, under the play of *invisible* light. Sir George Stokes showed in 1852 that the impinging rays have their undulations actually lengthened by the action of such kinds of matter, so as to become degraded in the spectrum, and thus brought within the range of vision.

The Herschelian theory of the sun was adopted, and long retained by Sir John. He believed in a cool, solid interior globe sheltered by a succession of aërial envelopes, rent, locally and temporarily, by tornadoes of fire. The presence of inhabitants on the globe so circumstanced was credible to him, although he abstained from dwelling upon the advantages of their state. He carefully followed, however, the progress of solar science, and in 1864 explained his altered views in the *Quarterly Journal of Science*. He now regarded the sun as a wholly gaseous mass—a conclusion in which he was anticipated only by Father Secchi. He added that it must be largely composed of matter kept in an intermediate condition between liquid and vaporous by "high temperature and enormous pressure." The spot-period, he suggested, might be that of a revolving meteoric ring with condensations.

He was vividly interested in the "willow-leaf" controversy, raised in 1862 by Nasmyth's misinterpreted observations. The objects seen were simply Sir William Herschel's "nodules"—the luminous elements of the sun, held by Sir John in 1867 "to be permanently solid matter, having that sort of fibrous or filamentous structure which fits them, when juxtaposed by drifting about, and jostling one against another, to collect in flocks as *flue* does in a room." He concluded with the remarkable assertion that the sun has no real surface, "the density diminishing from that below the photosphere to *nil* in the higher regions, where the pressure is *nil*."

Herschel's "Cape Observations" stands alone in astronomical literature for the wide and permanent interest of its contents. They are exceedingly various. Chapters on Halley's Comet, on Sun-spots, the

Satellites of Saturn, Astrometry, the Constitution of the Southern Galaxy, are associated with discussions on the nature and distribution of nebulæ, with monographs of two, and incidental notes on many of these enigmatical objects. The volume is illustrated with over sixty beautiful steel engravings of nebulæ and clusters, of sun-spots, and of the comet.

The speculations it includes regarding the nature of nebulæ, deserve even now to be remembered. Sir John was, at the outset, an unwavering adherent of the theory developed by his father in 1811. They were composed, he held in 1825, of a " self-luminous, or phosphorescent material substance, in a highly dilated or gaseous state, but gradually subsiding, by the mutual gravitation of its molecules, into stars and sidereal systems." His personal experience, however, ran counter to this view. In 1833 he had become convinced that a nebula is, in general, " nothing more than a cluster of discrete stars."

The successful resolution into stars, with the great Parsonstown specula, of many nebulæ until then called irresolvable, carried him still further in the same direction. To him, as to other thinkers, the presence in space of a self-luminous cosmic fluid became more than doubtful. In his Presidential Address to the British Association in 1845, he dwelt with enthusiasm on the completion of the Rosse reflector—" an achievement of such magnitude, that I want words to express my admiration of it." He regarded " as one of the grand fields open for discovery with such an instrument, those marvellous and mysterious bodies, or systems of bodies, the nebulæ." Their frequent resolution, actual or indicated, with increased optical power, led him to attribute recalcitrance in this respect to the smallness and closeness of the stars of which they consist; he held them, in short, to be "optically, and not physically, nebulous."

A new consideration was thus introduced into discussions on nebulæ. The whole burthen of accounting for their varieties in telescopic aspect need no longer be thrown upon differences of remoteness; diversities in the size and closeness of nebular *molecules* would answer the same purpose. So that pulverulent agglomerations, it was thought, might pass by insensible gradations into collections of truly sunlike bodies. All distinction between nebulæ and clusters was then abolished, the members of both classes consisting, like the sun's photosphere, of shining granules, supported in an obscure medium, varying in real magnitude from *floccules* to great globes, while each vast compound body rotated *en masse* on an axis. Whatever the merits of this scheme, it at least harmonises with the now prevalent opinion that nebulæ and clusters belong to one unbroken cosmical series. "They are divided," Mr. Cowper Ranyard wrote in 1893, "by no hard and fast line. The larger nebulæ may be described as groups of stars surrounded by bright nebulosity, and star-clusters as groups of stars surrounded by faint nebulosity."

Herschel's assimilation of nebulæ to clusters was not meant to apply to "those extraordinary objects resembling the wisps and curls of a cirrous cloud," which confront the astronomer in Orion, Argo, and elsewhere. "The wildest imagination," he said, "can conceive nothing more capricious than their forms. With their resolution," he averred, "and that of elliptic nebulæ, the idea of a nebulous matter, in the nature of a shining fluid or condensible gas, would cease to derive any support from observation." He, in fact, discarded it absolutely on the deceptive analysis into stars at Parsonstown and Harvard College of the Orion and Andromeda nebulæ. The discredited hypothesis was nevertheless triumphantly reinstated by Dr. Huggins's spectroscopic observations in 1864.

One-third of the whole nebular contents of the heavens Herschel found to be collected into a broad, irregular patch, the central point of which in Virgo coincides almost precisely with the northern pole of the Milky Way. He compared it to a canopy surmounting the galactic zone. In the other hemi-sphere the arrangement, although less distinctly, characterised, is on the same general plan. Plainly, then, nebular distribution has an opposite corre-spondence with stellar distribution, and the two partial systems are complementary one to another. Herschel, however, contented himself with the some-what ambiguous statement that "the nebulous system is distinct from the sidereal, though involving and, to a certain extent, intermixed with it."

His verdict as to the ground-plan of the sidereal edifice might be summed up in the phrase, "Not a stratum, but an annulus," our own situation being in a relatively vacant interior space. Hence, the sun belongs, not to the Milky Way proper—as it should on the stratum theory—but to the system of which the Milky Way forms part. This conclusion was in itself a distinct advance towards the solution of an exorbit-antly difficult problem. The grand question as to the remoteness of the star-clouds in that gleaming sky-girdle was definitely raised by it; and the question is not, in the nature of things, unanswerable. Herschel's annulus was not a neat structure with a cylindrical

section, but "a flat ring, or some other re-entering
form of immense and irregular breadth and thickness."
It is cloven over one-third of its circumference; it is
interrupted by huge chasms; it is bent, and shattered
and broken, and probably set with tentacular appendages, giving rise, by their foreshortening, to very
complex visual effects. All of which modifying circumstances Herschel implicitly recognised. He was the
first to gather any direct intimations of the existence
of that "solar cluster" which, guessed at by the elder
Herschel, has of late assumed a sort of elusive reality.
A zone of bright stars, including those of Orion, Canis
Major, the Ship, the Cross, and the Centaur, struck
him at once as a conspicuous feature in the scenery of
the southern heavens. Its aspect led him to "suspect
that our nearest neighbours in the sidereal system
form part of a subordinate sheet, or stratum," inclined
at an angle of twenty degrees to the plane of the
Milky Way. To Dr. Gould at Cordoba, in 1879, "few
celestial phenomena" appeared "more palpable" than
this projected star-belt; and, since it traces out a great
circle on the sphere, the sun must be placed within it,
and pretty accurately in its plane; yet the difficulty
of associating it intimately with our particular star
seems all but insurmountable.

Herschel's minor and occasional writings were
neither few nor unimportant. He contributed articles
on "Isoperimetrical Problems" and "Mathematics"
to Brewster's *Edinburgh Cyclopædia*, and on "Meteorology," "Physical Geography," and "The Telescope,"
to the eighth edition of the *Encyclopædia Britannica*.
These last were printed separately as well. He edited
in 1849 the Admiralty "Manual of Scientific Inquiry,"
and criticised in the *Edinburgh* and *Quarterly*

Reviews Mrs. Somerville's "Mechanism of the Heavens," Whewell's "History of the Inductive Sciences," Humboldt's "Kosmos," and Quetelet's "Theory of Probabilities." His addresses as President of the Royal Astronomical Society were models of their kind, and the same might be said of his memoirs of Baily and Bessel in the "Monthly Notices." Most of them were collected in 1857, with his review articles, into a volume of "Essays;" and his attractive "Familiar Lectures on Scientific Subjects," published in 1867, gave permanence to some popular discourses delivered in the school-house of Hawkhurst, as well as to articles from *Good Words* on Light and other subjects. No less than 152 papers by him are included in scientific repertories.

He had a considerable faculty for translating poetry, and its exercise made one of his favourite recreations. Having adopted the literal theory of the art, he kept strictly to the original metres, and thus fettered, got over the ground with more grace and ease than might have been expected. His first attempt with English hexameters was in a version of Schiller's "Walk," privately printed in 1842. He had come to love the poem through its association in his mind with a favourite stroll up the side of Table Mountain; and a translation of it in the *Edinburgh Review* leaving, as he thought, something to be desired, he tried his hand, and distributed the result "among his friends as his Christmas sugar plum." The various acknowledgments made an amusing collection. One lady said that she "found it difficult to get into the step of the *Walk*." Another correspondent declared that the *Walk* had got into a *Run* through ceaseless borrowing. A third qualified his encomium upon the ideas by adding, "To the *verse* I am *averse.*" Joanna Baillie, however, and her sister were delighted with both the substance and form of the poem, and it was included among Whewell's " English Hexameter Translations " in 1847.

His success encouraged him, after twenty years, to undertake an indefinitely more difficult task.   Pope's Iliad he described happily as " a magnificent adumbration " of the original ; but he aimed rather at producing a " fac-simile," in

" Hexameters no worse than daring Germany gave us."

His version should come as near as he could bring it to a photograph of a grand piece of architecture; and as a measure of its fidelity, he printed in italics all the words *not* in the text.   Whewell remarked that it was " curious to see how few he had managed to make them," and preferred his translation to any other with which he was acquainted.   But English hexameters were a hobby of the Master of Trinity, who accordingly viewed with partiality what Tennyson called the " burlesque barbarous experiment " of thus lamely rendering " the strong-wing'd music of Homer."

De Morgan, too, was one of the " averse."   " Many thanks for the hexameters," he wrote, on receiving an instalment of the Collingwood Iliad; " they are as good as they can be, but all the logic in the world does not make me feel them to be English metre, and they give satisfaction only by reminding one of the Greek : just as, mark you, a flute-player—which I have been these forty-five years—only plays Haydn and Mozart because he has the assistance of the orchestral accompaniment which arises in his head with the melody. The hexameter, it is clear, does not fix itself in the

o

popular mind. The popular mind knows neither quantity nor accent, but that which is to last bites its own way in, without any effort."

Yet Herschel's translation is not without merit. It is disfigured neither by affectation nor by magniloquence, and it catches here and there something of the greatness of the unapproached original. Let us take two specimens ; this from the " Shield of Achilles " :—

" There he depicted the earth, and the canopied sky, and the
    ocean ;
There the unwearied sun, and the full-orb'd moon in their
    courses.
All the configured stars, which gem the circuit of heaven,
Pleiads and Hyads were there, and the giant force of Orion.
There the revolving Bear, which the Wain they call, was en-
    sculptured,
Circling on high, and in all its course regarding Orion;
Sole of the starry train which refuses to bathe in the Ocean."

The next likewise appeals to the astronomer. It is the famous simile from the end of the Eighth Book :—

" As when around the glowing moon resplendent in ether,
Shines forth the heavenly host, and the air reposes in stillness;
Gleams every pointed rock, stands forth each buttress in pro-
    spect ;
Shimmers each woodland vale ; and from realms of unspeak-
    able glory
Op'ning, the stars are revealed ; and the heart of the shepherd
    rejoices.
Such, and so many the fires, by the Trojans kindled, illumined
Eddying Xanthus' stream, and the ships, and the walls of the
    city."

Sir John Herschel corresponded with Mr. Proctor, during the last two years of his life, on the subject

of sidereal construction; and his replies to the arguments put before him show that his mind retained,
even then, its openness and flexibility. He had none
of the contempt for speculative excursions which
sometimes walls up the thinking-powers of observers.
" In the midst of so much darkness," he held that
" we ought to open our eyes as wide as possible to any
glimpse of light, and utilise whatever twilight may be
accorded us, to make out, though but indistinctly, the
forms that surround us."   " *Hypotheses fingo* in this
style of our knowledge," he went on, " is quite as good
a motto as Newton's *non fingo*—provided always they
be not hypotheses as to modes of physical action for
which experience gives no warrant."   And again :
" We may—indeed, must—form theories as we go
along ; and they serve as guides for inquiry, or
suggestions of things to inquire ; but as yet we must
hold them rather loosely, and for many years to come
keep looking out for side-lights."

These were his last words on the philosophy of
discovery: and they constituted his last advice to
scientific inquirers. But, good as were his precepts,
his example was better. There was no discrepancy
between his work and his thought. Both combined
to inculcate aloofness from prejudice, readiness of
conviction in unequivocal circumstances, suspension
of judgment in dubious ones, and in all, candour,
sobriety, and an earnest seeking for truth.

INDEX.

Printed in the United States
By Bookmasters